The Role of Integrity in the Governance of the Commons

Laura Westra • Janice Gray •
Franz-Theo Gottwald
Editors

The Role of Integrity in the Governance of the Commons

Governance, Ecology, Law, Ethics

Springer

Editors
Laura Westra
Maple, Ontario
Canada

Janice Gray
Faculty of Law
University of New South Wales
Sydney, New South Wales
Australia

Franz-Theo Gottwald
Schweisfurth-Stiftung
Munich
Germany

ISBN 978-3-319-54391-8 ISBN 978-3-319-54392-5 (eBook)
DOI 10.1007/978-3-319-54392-5

Library of Congress Control Number: 2017942767

Printed on acid-free paper

This Springer imprint is published by Springer Nature
The registered company is Springer International Publishing AG
The registered company address is: Gewerbestrasse 11, 6330 Cham, Switzerland

Preface

The 24th meeting of the Global Ecological Integrity Group took place in Munich, Germany, under the auspices of Franz-Theo Gottwald, a long-time member. It was a particularly appropriate location, given the emphasis on green spaces and healthy food that pervades that city. Hence it seemed right to use a German Press for our collection, perhaps in order to prolong the memory of that beautiful city.

We were extremely lucky to have Peter H. Sand to open the conference, a scholar no doubt cited by most of us, but not met by many, including the editors. His chapter traces the movement of international law towards the acknowledgment of the global commons (now accepted by both the UNESCO World Heritage and the FAO Plant Genes Regimes), to be "within the territorial jurisdiction of States". As well, "proprietary sovereign rights" can now be limited by norms such that the states involved may be "accountable as trustees". Such developments give hope, as they represent clear steps towards Earth Governance.

Franz-Theo Gottwald's chapter presents a scathing critique of synthetic biology, a discipline which "creates self-replicating organisms destined to be released into the environment", with enormous security risks, which are not properly addressed by either their producers, distributors, or the appropriate governmental or legal agencies, intended for the protection of the public. Biosafety is not pursued in ecology, agriculture, medicine, and several other fields. Further, the precautionary principle is not applied. Thus there is no "ethical protocol on integrity and the preservation of life-forms".

Agnes Michelot and A. Aseeva address the question of justice regarding environmental issues and the need to appreciate and protect value in ecology, not through the commodification of "ecosystem services", but through "ecological solidarity". The latter is based on the "natural spatial and temporal interdependence among entire ecosystems". This approach fosters relational justice, thus offering a way beyond both "ecocentric and anthropocentric ethics".

In the last chapter of Part I, Klaus Bosselmann returns to the challenge of the "global commons", as it emerges against the background of an ongoing "democratic vacuum at the global level", and the ever-increasing power of multinational corporations, the main characteristic of globalisation. This situation renders urgent the need to reclaim the Earth for global citizens, through the concept of state trusteeship.

The second part opens with Janice Gray's discussion of water law and governance in which she observes that while the high seas are classified as a global commons, terrestrial waters have not yet been so classified. They continue to be governed largely by domestic law with some limited incursions of international law and some examples of international river basin agreements. This position leads her to emphasise the importance "of getting domestic water law and governance right" particularly when the over-arching guiding principles of international law, such as the "common heritage of humankind" principle, are not necessarily part of domestic, terrestrial water law. She argues that in the Australian context, public interest litigation is an important tool for strengthening domestic water law and governance. However she notes that a range of factors impact on the ability to bring public interest suits. Those factors include justiciability, cost and standing, for example. Gray then analyses two legislative amendments: one which would abolish representative standing for environmental organisations, and another which introduces strong deterrents to protest. She concludes that these amendments could impact negatively on the use of public interest litigation to enhance water law and governance. They certainly go to the heart of effective, robust and participatory democracy.

In the chapter "The Water Crisis in Flint, Michigan: Profitability, Cost-Effectiveness, and Depriving People of Water", Joseph W. Dellapenna addresses the right to water, through a discussion and analysis of the water crisis which took place in 2014–2015 in Flint, Michigan. This example shows clearly the conflict between the "quest for profitability" and "cost-effectiveness", and public health. The progression from the change in water provenance, the neglect of older equipment and in general the avoidance of controls for the protection of the public resulted in lead exposure for children, in elevated e-coli levels resulting in disease and death, and in an outbreak of legionnaire's disease. Dellapenna analyses the multiple factors involved in the crisis, including racial and social issues.

Katy Kintzele Gwiazdon discusses the contentious geopolitical issues in the South China Sea associated with China's maritime claims in this region and its associated conduct which has impacted negatively on coral, endangered species and fisheries to name but a few areas. She also cites examples of aggressive island building on sites which were once only single uninhabitable rocks visible at high tide. Such island building is, she suggests, designed to bolster China's maritime claims and extend its territory into resource-rich zones. Gwiazdon employs a human security lens to provide context to the discussion and she explores the components that foster human security as well as the way in which those components correlate to the relationship and resources in the South China Sea. Gwiazdon is concerned to demonstrate how a cooperative resolution of the present tensions may be effectuated and she emphasises the importance of doing so in order to protect the environment from more immeasurable harm.

Part III starts with Donald A. Brown, who analyses and discusses the damage caused by the disinformation campaigns waged for decades against the scientific facts of climate change. Those campaigns have not only denied evidence of climate change but also the role of human causality in that change. They argue that "more harm than good" would be caused by reducing greenhouse gasses. Brown details the huge amounts of money spent to fund numerous groups supporting misleading and false claims through the media, particularly through the work of corporate funded Think-Tanks and other groups which jointly have been responsible for at least a 50-year delay in the steps required to reduce the threat of climate change.

In the chapter "The Projection of Global and Regional Climate Change Models into Selected Ecosystem Functions and Services (Case Study Czech Republic)", Pavel Cudlín discusses several global and regional climate change models up to 2000, including how the emission scenarios of IPCC RCP 4.5 and 8.5 were applied to selected ecosystem functions (e.g. production function) and services (e.g. carbon sequestration, habitat services) at different scale levels (from small catchment to whole republic) in the Czech Republic. He observes that the Land Change Modeller, InVEST and Globio models were used for prediction of land use/land cover and the ecosystem functions/services. He notes that his prediction of the impacts of climate factor changes on the landscape up until 2000 indicates the extensive decrease in important ecosystem function performance and ecosystem service provision in the second half of last century. These changes, including gradually accepted mitigation and adaptation measures, will, he concludes, result in a substantial ecosystem service trade-off and continuous biodiversity loss.

Eva Cudlínová (tenth chapter) asks the question whether the new "bio-economy" may help mitigate climate change. Bio-economy has been discussed in both political and legal documents as well as in scientific works. However, although it is promoted as a novel step forward towards climate change mitigation, biomass production is the source of many other problems. Even the possibility of "replacing fossil fuels with bio-energy" may not reduce carbon emissions, especially as "bio-energy crops displace forests and grasslands". This chapter also raises the question of land availability, noting "land-grabbing" produces grave harms in Africa and Asia.

Part IV starts with a discussion of sustainable development, by Massimiliano Montini and Francesca Volpe. They are interested in the role and status of the concept of sustainable development and noting that international law is at the crossroads between economic development, social development and environmental protection, wonder what this will mean for sustainable development. Will the concept be revitalised or like the protagonist in the film, "Sunset Boulevarde", will it slide into oblivion? Montini and Volpe isolate three independent yet concatenate events which they believe might exercise influence on shaping the principle's future. Those events are (a) the publication of Pope Francis's Encyclical Letter Laudato Sì, (b) the adoption by the UN General Assembly of the Sustainable Development Goals and the related 2030 Agenda for Sustainable Development and (c) the conclusion of the Paris Agreement on Climate Change. Montini and Volpe conclude that the role sustainable development plays in the near future will not depend merely on the independent legacy of the three events described above, but rather on their systemic integration and alignment.

In the chapter "The Ecological Catastrophe: The Political-Economic Caste as the Origin and Cause of Environmental Destruction and the Pre-announced Democratic Disaster", Donato Bergandi addresses the ecological crisis which he terms a "dystopian ecological catastrophe", as it enriches a few but is the cause of pollution and environmental destruction for the many". The paradigm of sustainable development has emerged without "calling into question the economic production systems". Bergandi cites the utilitarianism of both Mill and Bentham, who acknowledge the dangers of dominant classes and influence governments to promote their own interests against the good of the whole community. That is why the "current system of representative democracy is completely disconnected from…the pursuit of the common good". Hence he argues the present environmental situation should be accepted as a moral challenge for humanity.

In the chapter "Ecological Integrity in the Anthropocene: Lessons for Law from Ecological Restoration and Beyond", Geoffrey Garver argues that "downsizing and stabilization of the economy is urgently needed to reverse global ecological trends". The human relationship to Earth must acknowledge and respect the role that each organism has to play, both human and nonhuman, in order to achieve a "human inclusive ecocentric paradigm". Ecological integrity and "related notions" remain integral to an ethic appropriate to the anthropocene era.

Part V explores the human responsibility for the current crises. In the chapter "Addressing the Problem of Conflict-of-Interest and Moneyed Influence in Public Health: Some Case Studies", Colin L. Soskolne examines the problem of conflicts of interest between "experts" and the public interest, as the former are often supported and promoted by interested parties. Epidemiology is "a most critical science used to inform public health policy". When "moneyed influence" infiltrates science and the literature upon which public policy is founded, the damages to the health and the life of the public are incalculable.

In the chapter "Ethics and Pesticides: The Precautionary Principle as Illustrated by Glyphosate", Josef Unterweger moves from theory and general legal and moral assessments to legal practices concerning genetically modified organisms, and glyphosate, perhaps the most infamous product of the giant producer of both GMOs and pesticides, Monsanto. The difficulties of bringing to justice a major corporation, whose products are known and proven carcinogenic, and the effects of which adversely affect human beings from conception to old age, are documented and discussed.

In the chapter "Laudato Sì and the Christian Ecological Utopia", Philippe Crabbé discusses the 2015 Papal Encyclical "Laudato Sì" in some detail. Crabbé starts by tracing the historical antecedents of Pope Francis's position, as most of the concepts and arguments found in that document, Crabbé argues, have been discussed by earlier Church authorities. Nevertheless most of the concepts and arguments that animate Laudato Sì have been discussed and analysed by members of the Global Ecological Integrity Group for two years and have been declared in the Earth Charter as well. In contrast, we should note that the arguments advanced in the Encyclical are much closer to earlier Church authorities than they are to recent environmental ethics.

Peter Venton also examines the Papal Encyclical on ecology in the chapter "Pope Francis's Ethics for Democratic Capitalism and the Common Good". He observes that in the Encyclical Pope Francis appealed for a new dialogue with people about shaping the future of our planet. Venton sees the Encyclical as constituting a vigorous attack on the ethics, politics and the economics of "neo liberal" capitalism and he argues that implicit in the encyclical's critique are proposals for "democratic capitalism" to replace the neo-liberal version of capitalism. He explains that democratic capitalism is about three dynamic systems converging as one: a democratic polity, a capitalist economic system based on markets and incentives, and a moral-cultural system which is pluralistic and, in the largest sense, liberal. Venton concludes that the concept of democratic capitalism matches most of Pope Francis's ethics and his vision of the common good for humanity.

Finally in the chapter "Natural Catastrophes and Forms of Catastrophism. A New Ethical and Moral Framework Leading Towards the 'Responsible Catastrophism Model'", Marco Ettore Grasso proposes several ways of dealing with the presently growing and rapidly peaking environmental catastrophes. He argues that we need to start by acknowledging our human limitations and our vulnerability in the face of global disasters, such as climate change. We need to study the causes of such disasters and learn to cooperate in order to prevent their arrival as much as possible. Finally, we need to cultivate solidarity among humans in order to acknowledge with Hans Jonas the principle of responsibility, more necessary than ever at his time.

We commend this book to the reader and hope that it raises interesting and challenging issues about the commons, governance, ecology, law and ethics.

Maple, ON, Canada — Laura Westra
Sydney, NSW, Australia — Janice Gray

Contents

Part I
Governance for the Commons

Accountability for the Commons: Reconsiderations

Peter H. Sand

1 Dedication

Let me start with a triple *caveat*. First, English is *not* my mother tongue. And while most of us now use and misuse William Shakespeare's language in the way scholars formerly used Latin—as a means of universal communication with scholars of other nations, including the anglophones but not them alone—a native Bavarian speaking to you in English may be excused for what Dutch Supreme Court Justice Huibert Drion once compared to "the kind of frustration suffered by the person who attends a formal dinner in borrowed clothes which he knows do not fit too well" (Drion 1954, p. vi).

Secondly, I am a newcomer to your group. And while I have long followed your work with keen interest—especially the tireless efforts of Laura Westra to raise the concept of 'global ecological integrity' to the level of recognition it deserves (Westra 1994, 2016)—I hope you will bear with me if I am not fully conversant with the kind of discourse and terminology which the insiders among you may take for granted.

Thirdly—and that is a real handicap—I happen to be an international lawyer. Even though I shall try to be as interdisciplinary as I can, my *déformation professionnelle* will inevitably shine through as I proceed. And since we are not very far here from the Law Faculty of Munich University, let me take this opportunity to dedicate my presentation today to the memory of someone whom many of my colleagues consider as the founding father of International Environmental Law as an academic discipline in Germany, and perhaps even worldwide: Karl Alexander Neumeyer, who taught international law at the University of Munich from 1901 to 1933 (Sand 2012, p. 185; Sand 2015, p. vii).

P.H. Sand (✉)
Institute of International Law, University of Munich, Munich, Germany
e-mail: p.sand@jura.uni-muenchen.de

L. Westra et al. (eds.), *The Role of Integrity in the Governance of the Commons*,
DOI 10.1007/978-3-319-54392-5_1

Of course, the term 'environmental law' (*Umweltrecht*) did not even exist in German legal language at that time. Yet, Neumeyer's monumental four-volume treatise on what he called 'international administrative law' (*Internationales Verwaltungsrecht*) assembled and analysed a unique compendium of contemporary legal source materials that would indeed qualify today as typical 'transnational environmental law'. Chapter 8 in volume 2 of his treatise, first published in 1922, was thus titled 'natural resources and products' (*Naturkräfte und Naturerzeugnisse*; Neumeyer 1922). It dealt with internationally shared water resources and water power; the transboundary regulation of mineral resources, agriculture, forestry, hunting and fishing; and the management and conservation of marine living resources.

Neumeyer tragically did not live to see his pioneering work generally accepted. He was of Jewish ancestry; when the Nazi regime took over in Germany, he was forced into retirement, and barred from working with the Hague Academy of International Law (where he had first lectured in 1923) and the *Institut de Droit International* (which had elected him to full membership in 1926). Ultimately, when he was notified of the impending eviction from his family home and the confiscation of his precious private library, he and his wife decided to commit suicide in July 1941—almost exactly 75 years ago (Morgenthau 1941; Wehberg 1941; Gutzwiller 1947; Vogel 1970; Vogel 2001; von Breitenbuch 2013). There is a memorial tablet for them outside their former home (at Königin-Str. 35a, just around the corner from here); and in 2008, the Munich Law Faculty (whose dean Karl Neumeyer had been in 1931–1932) named the building that houses its Institute of International Law (which he had helped to create, at Veterinär-Str. 5, close by) in his honour and memory.

Let me now turn to the substance of my chosen topic, 'accountability for the commons'. There has been an extraordinary renaissance of the commons debate in recent years, both at the national and the international level, and over a wide range of disciplines—all across economics, political science, sociology, anthropology, ecology, ethics, and the law (Buck 1998; Vogler 2012; Wall 2014); and all the way from Garrett Hardin's classic essay (Hardin 1968) and the work of Nobel Laureate Lin Ostrom (Ostrom 1990),[1] to the valiant drafting efforts of Stefano Rodotà and his *benecomunisti* (Rodotà 2013; Mattei 2015; Capra and Mattei 2015, pp. 149–168).

To narrow down that somewhat intimidating spectrum of scholarship, however, I propose to focus on two issues of particular concern to me:

- How do the Earth's global commons fit into the contemporary world of sovereign States? and

[1]The German synonym of Ostrom's concept of the 'commons' is either the medieval term *Allmende* (Ostrom 1999, reminiscent also of Scandinavian *allemansrätt*) or in modern usage *Gemeingüter* ('common goods'; Ostrom 2011). By contrast, the interdisciplinary research scope of the Bonn-based *Max-Planck-Institut zur Erforschung von Gemeinschaftsgütern* (translated as 'collective goods') includes both environmental and economic/financial aspects of governance.

- How can sovereign States be held accountable for the ecologically sound management of our global commons?

2 Global Commons and Sovereign Prerogatives

The standard legal textbook definition of the global commons is invariably a negative one: i.e., areas or resources that are *not* subject to the exclusive territorial sovereignty of States (Kish 1973; Wolfrum 1984; Cleveland 1990; Tomuschat 1993; Stone 1993; Durner 2001; Joyner 2001), such as the high seas, the seabed below them and the atmosphere above them; Antarctica[2]; outer space; and possibly the electromagnetic radio-spectrum and the geostationary satellite orbit (Kiss 1982, pp. 145–151, 157–160).[3] In a way, that spatial perspective reflects the prevailing 'territorial obsession' of international lawyers ironically diagnosed by Scelle (1958), or the less benign 'spatial ontology' postulated by Schmitt (1997, Minca and Rowan 2015); or—*magari*—the 'territorial imperative' which could well be part of our ancient genetic heritage from the animal kingdom (Ardrey 1966; Khan 2012). Be that as it may, the fact remains that even in domains long identified as *res communes omnium*, national governments have already secured enclosures (e.g., via the 'sovereign rights' of coastal States, under the UN Convention on the Law of the Sea, in the 200-mile exclusive economic zone and up to a 350-mile continental shelf margin); maintained old sovereignty claims (e.g., the temporarily 'frozen' territorial claims by seven States in Antarctica, some of which are overlapping); or raised potential new issues of access and benefit (e.g., commercial exploitation of mineral resources on celestial bodies).[4]

[2]Schrijver (2016) includes both polar regions in this context, though noting the continuing (and partly conflicting) territorial claims of the four Arctic countries.

[3]Article 44(2) of the ITU Constitution 1992 recognizes radio frequencies and the geostationary-satellite orbit as "limited natural resources" to which all countries shall have equitable access; see Ryan (2004); Lyall (2011), pp. 127–191, 245–256; von Schorlemer (2012), p. 826.

[4]The U.S. *Space Resource Exploration and Utilization Act* 2015, while affirming that "by the enactment of this Act, the United States does *not* thereby assert sovereignty or exclusive rights of jurisdiction over, or the ownership of, any celestial body" [emphasis added], goes on to stipulate that "a United States citizen engaged in commercial recovery of an asteroid resource or a space resource under this chapter shall be entitled to any asteroid resource or space resource obtained, including to possess, own, transport, use and sell the asteroid resource or space resource obtained in accordance with applicable law, including the international obligations of the United States". Similar legislation is now under preparation in Luxembourg (host country of the *Société Européenne des Satellites* and several other aerospace companies), according to a Government press release of 3 February 2016 ("development of a legal and regulatory framework confirming certainty about the future ownership of minerals extracted in space from Near Earth Objects such as asteroids", "in full consideration of international law" and "without damaging natural habitats"). For background see Lyall and Larsen (2009), pp. 175–197; Lee (2012), Lewis (2014), MacWhorter (2016).

True enough, the exercise of State powers in those domains has also been tempered by concepts of international community interest, such as 'common heritage' (Taylor and Stroud 2013), and 'common concern'.[5] By and large, however, powerful States have persistently and successfully defended their customary sovereign prerogatives against most attempts at reining them back (Milun 2011).

A pertinent recent example is the ongoing discussion on protection of the atmosphere in the UN International Law Commission (ILC). Lawyers, economists and scientists alike have long categorized the atmosphere as 'true global commons' (Obama et al. 1991, p. 1536; Soroos 1997; Soroos 1998; Vogler 2001; Harrison and Matson 2001; Wustlich 2003; Halfmann 2012; Coghill et al. 2012; Everard et al. 2013). After preliminary discussions in 2011–2012, the ILC inscribed the topic on its programme of work in 2013 and appointed Professor Shinya Murase (Sophia University/Tokyo) as Special Rapporteur. In a first syllabus, he had boldly envisaged "a comprehensive set of draft articles for a framework convention on the protection of the atmosphere" (Murase 2011, p. 317; and Murase 2012), along the lines of part XII of the 1982 UN Convention on the Law of the Sea (protection and preservation of the marine environment).

From the beginning, however, there was considerable opposition to this approach on the part of Commission members from the 'Big Five' (the permanent member countries of the UN Security Council), whose diplomatic representatives had already criticized Murase's proposal during debates in the General Assembly's Sixth Committee in 2011, suggesting either that it was "too technical" for the ILC, or that there was no need for codification in this field at all (UNGA 2011). Even though the Rapporteur went out of his way to reaffirm the principle of sovereignty of States over their national airspace, the sheer prospect that the proposed draft articles would also apply to "certain activities on the ground within a State's territorial jurisdiction" (Murase 2011, p. 318) was evidently enough to raise instant political alarm among what Philip Allott calls "the international *Hofmafia*" of lawyer-diplomats (Allott 2002, p. 384, borrowing a term from Wheatcroft 1996, p. 248; see also Koskenniemi 2005, p. 336). As a result, after non-public deliberations in the ILC Planning Committee, the Commission adopted a highly restrictive 'Understanding', reading (ILC 2013, p. 115, para. 168):

(a) Work on this topic will proceed in a manner so as not to interfere with relevant political negotiations, including on climate change, ozone depletion, and long-range transboundary air pollution. The topic will not deal with, but is also without prejudice to, questions such as liability of States and their nationals, the polluter-pays-principle, the precautionary principle, common but differentiated

[5]Note, however, that the 'common concern of humankind' acknowledged in the preamble of the UNFCCC (1992) (reaffirmed in the preamble of the 2015 Paris Agreement) does *not* refer to the atmosphere or climate as such, but to "change in the Earth's climate and its adverse effects" (Brunnée 2007, p. 565). By contrast, the IUCN Draft Covenant (IUCN 2015, Article 3) more generally refers to "the global environment" as "a common concern of humanity".

responsibilities, and the transfer of funds and technology to developing countries, including intellectual property rights.
(b) The topic will also not deal with specific substances, such as black carbon, tropospheric ozone, and other dual-impact substances, which are the subject of negotiations among States. The project will not seek to "fill" the gaps in the treaty regimes.
(c) Questions relating to outer space, including its delimitation, are not part of the topic.
(d) The outcome of the work on the topic will be draft guidelines that do not seek to impose on current treaty regimes legal rules or legal principles not already contained therein.
(e) The Special Rapporteur's Reports would be based on this understanding.

These perplexing amputations of the Rapporteur's mandate prompted consternation and severe criticism not only by external academic commentators (Plakokefalos 2013; Lode et al. 2016, p. 32; Sand and Wiener 2016, pp. 208–216), but also from ILC members, who did not hesitate to characterize the unprecedented 'understanding' as a "disgrace to the Commission" (Candioti 2014, p. 7), wondering whether it had been "purposely designed to bog down the work on the topic" (Peter 2015, p. 12). Yet, at the 67th session in 2015, the ILC Drafting Committee went one step further by insisting that the terms of the 'understanding' be moved from the preamble to the operative body of the draft guidelines (ILC 2015, p. 33). Conversely, the Committee rejected Murase's proposal to proclaim the protection of the atmosphere a "common concern of humankind" in draft guideline 3, and instead settled for the seemingly innocuous formula "pressing concern of the international community as a whole" in a mere preambular paragraph, explaining the expression "as a factual statement, and not a normative statement" (ILC 2015, pp. 22–26). Drafting work will continue at the 69th ILC session in 2017, though, and hope remains that the project—even with (or in spite of) its torso of a mandate—may still be able at least to redress some of the dysfunctions of the fragmented 'regime complex' (Keohane and Victor 2011) of the global atmospheric commons.

3 Towards Public Trusteeship for the Commons?

But let us come back to Earth. As Louis Sohn at Harvard Law School used to say, international lawyers need to be like giraffes: They may have their heads in the clouds, but they should have their feet on the ground.[6] Hence, for an international legal regime to be viable and effective, it would also have to be actionable in court.

[6]Which prompted *Philip Allott* in turn to declare himself a "legless giraffe" (Scobbie 2005, p. 313).

Yet, as the ILC experience demonstrates, the chances of global community interests being defended by way of traditional inter-state remedies—such as adversarial litigation before international tribunals, or countermeasures by States not directly affected—are remote. Well before its current tergiversations over protection of the atmosphere, the International Law Commission had relegated the enforcement of *erga omnes* obligations to "the further development of international law", in a controversial savings clause added to its 2001 Draft Articles on State Responsibility (ILC 2001, p. 355). And in international judicial proceedings, States only rarely take the steps required to formally invoke the law of state responsibility against other States' breaches of obligations owed to the community as a whole (Brunnée 2005, p. 21; Tams 2011, pp. 383–388),—to the point where critical observers have described the *erga omnes* construct as "the wishful thinking of publicists" (Rubin 1993, p. 172).

By default, then, the task of acting as 'guardians' of environmental community interests has fallen on non-state actors (Sands 1989, p. 417). In view of their lack of standing to sue before most existing international courts, however, NGOs can only operate within national judicial systems, or alternatively through the new 'non-adversarial' accountability mechanisms established by some multilateral environmental agreements (Pitea 2005; Epiney 2006; Treves et al. 2009) and multilateral financial institutions (Van Putten 2008, pp. 66–162).

Among recent examples of this type of civic litigation for protection of the commons are:

- a judgment by the UK Supreme Court in April 2015 (reaffirmed by the High Court in November 2016), declaring the British Government in breach of the European Union's Air Quality Directive for nitrogen dioxide (NO_2) emissions (*ClientEarth v. Department for the Environment, Food and Rural Affairs* 2015; Barritt 2015; Carrington 2016);
- a judgment by a civil district court in The Hague in June 2015, ordering the Dutch Government to reduce national annual greenhouse emissions by 25% by 2020 compared to 1990 levels (*Urgenda et al. v. Ministry of Infrastructure and the Environment* 2015; Peeters 2016);
- a judgment by the Massachusetts Supreme Court in May 2016, ordering the State Government to implement existing legislation for annual limits on greenhouse gas emissions (*Kain et al. v. Department of Environmental Protection* 2016; Wood and Woodward 2016, p. 645)[7];

[7]The case (remanded to the Superior Court for a new judgment) is part of a series of *Children's Atmospheric Trust* actions brought by public interest NGOs in the United States and in several other countries.

- and last week's decisions by an administrative tribunal here in Munich, ordering the Bavarian State Government to ensure compliance with the applicable EU air quality standards for NO_2 emissions in the city (especially from diesel car emissions) by June 2017 at the latest, under threat of an administrative fine of 10,000 Euros (*Deutsche Umwelthilfe* and *Verkehrsclub Deutschland v. Ministry of Environment and Consumer Protection* 2016).[8]

All these actions in court were brought by environmental NGOs, to hold governments accountable for their failure to protect an endangered common resource—the Earth's atmosphere. They may indeed be viewed as manifestations of 'public trusteeship for the commons' (Bosselmann 2015), a concept which can be traced back almost two millennia to Roman law.[9] In modern times, the concept underwent a remarkable metamorphosis, in the form of the 'public trust doctrine' as developed in contemporary environmental jurisprudence in the United States and a number of other countries, including India and South Africa (Razzaque 2001; Van der Schyff 2013; Wood 2013); acknowledged, inter alia, in a much-quoted separate opinion by Judge Weeramantry at the International Court of Justice, affirming a "principle of trusteeship for earth resources" (*Gabcikovo-Nagymaros case* 1997, p. 106).

In very simplified language, the doctrine means that

(a) certain natural resources—regardless of their allocation to public or private uses—are defined as part of an 'inalienable public trust';
(b) certain authorities are designated as 'public trustees' to guard those resources; and
(c) citizens, as 'beneficiaries of the trust', may invoke its terms to hold the trustees accountable and to obtain judicial protection against encroachments or impairments (Sand 2004, p. 49) (Fig. 1).

In an inter-temporal context (Brown Weiss 1989; Redgwell 1999), the beneficiaries also include future generations, as postulated more than 150 years ago by Karl Marx:

> Even society as a whole, a nation, or all contemporary societies taken together, are not owners of the Earth. They are merely its occupants, its users; and as diligent caretakers, must hand it down improved to subsequent generations. (Marx 1865)

In order to enforce the terms of the public trust against the trustees, therefore, the beneficiaries (present and future) need procedural safeguards, including actionable

[8]The cases are still subject to appeal; meanwhile, further initiatives are underway for class actions in Germany along the lines of the *Children's Atmospheric Trust* cases mentioned above.

[9]According to the Institutes II.1.1 (*de rerum divisione*) and the Digest I.8.2.1 of the *Corpus Iuris Civilis* of Emperor *Justinian I.* (533 A.D., which in turn were based on vol 3 of the Institutes of *Aelius Marcianus*, c. 220 A.D.), "surely by the law of nature, the atmosphere, watercourses, the sea and hence the seashores, are common to all" (*et quidem naturali iure omnium communia sunt illa: aer, aqua profluens, et mare, et per hoc litori maris*). English translations in Sanders (1903), p. 90; and Monro (1904), vol 1, pp. 39–40.

Fig. 1 International public trusteeship

rights to know, rights to be heard, and rights of standing to challenge governmental decisions (the 'three pillars' of the *Aarhus Convention* 1998; Ebbesson et al. 2014). Significantly perhaps, current atmospheric trust litigation is not primarily about monetary compensation. What the plaintiffs seek to obtain instead are declaratory judgments,[10] establishing the responsibility of governmental trustees for their management (or mismanagement) of public trust resources, through an accounting of the trust assets (e.g., in the form of air quality inventories and emission reduction plans); or injunctive relief, such as a denial of permits for activities harming the trust resources (Wood 2009, pp. 102, 114). Hence, as distinct from retrospective liability suits (Fitzmaurice 1996; Faure and Peeters 2011; Lord et al. 2011), the focus of public trusteeship—national or international—typically is on remedies *ex ante*, which may more appropriately be categorized as measures to ensure the trustees' continuous "legal accountability for the exercise of social power" (Allott 2001, p. 336).

That, however, raises an intriguing semantic issue. While 'accountability' is a household term in the jargon of public administration and political science (Mulgan 2000; Rached 2016; Keohane 2003, p. 154; Najam and Halle 2010; Baber and Bartlett 2016; Kramarz and Park 2016), the English-language legal triad of 'responsibility/accountability/liability' has no precise equivalent in a number of other legal tongues. For example, in the Romance languages (French, Italian, Spanish, Portuguese) all three concepts are rendered by a single polyvalent term (*responsabilité, responsabilità*); the same is apparently true for the legal vocabulary of Russian and other Slavic languages, whereas in Japanese and Hebrew, 'accountability' is rendered by simple transliteration of the original English word (Sinclair 1995; Richard 2011; Dubnick 2014). German legal usage does distinguish responsibility (*Verantwortung*) from liability (*Haftung*), but the equivalent of accountability is merely approximated by terms like *Rechenschaft* or *Zurechenbarkeit* (i.e., the duty to render accounts, etymologically close to antiquated English 'reckoning', Dutch *rekenschap*, or Swedish *räkenskap*). The dilemma is illustrated in the pioneering work of Hans Jonas, *Das Prinzip Verantwortung* (Jonas 1984a), which he himself (mis)translated into English as "the imperative of responsibility" (Jonas 1984b); yet, it is clear from the original text that what he meant was not necessarily responsibility in a legal sense, but something more akin to accountability.[11]

[10]As stated by the Massachusetts Supreme Court (*supra* note 7), "declaratory judgment is appropriate here".

[11]E.g., see p. 174 of the German version, where *Verantwortung* is defined as the precondition for *Rechenschaft*.

A similar Babylonian confusion obfuscates the very concept of international trusteeship, given that the common-law trust has no direct equivalent in traditional European civil law (Fratcher 1973; Waters 1995; Hansmann and Mattei 1998). Consequently, when Woodrow Wilson's terms "sacred trust of civilization" were inserted in Article 22 of the League of Nations Covenant in 1919 (from where they moved to Article 73 of the UN Charter in 1945), they were notoriously mistranslated as "*mission sacrée*" in the official French text, and as "*heilige Aufgabe /heiliger Auftrag*" in the German version (Jacobs 2004, pp. 82, 111; Matz 2005, pp. 50, 71). And when the United States submitted a draft "World Heritage Trust Convention" to UNESCO in 1972 (Train 1972; Meyer 1976, p. 48), the 'trust' term was subsequently deleted from the final text because it was considered untranslatable into French (Batisse and Bolla 2003, p. 17; Redgwell 2007, p. 268).[12] Alas therefore, an interdisciplinary perspective will also have to take into account the vicissitudes of comparative linguistics.

That did not, however, prevent the *World Heritage Convention* 1972 of the United Nations Educational, Scientific and Cultural Organization (UNESCO) from evolving towards an innovative legal regime that comes rather close to the idea of global public trusteeship (Kiss and Shelton 2007, p. 16; Benvenisti 2013, p. 329):

(a) world heritage sites are dedicated [as *corpus* of the trust] through nomination by a host state and acceptance of the nomination by the World Heritage Committee (WHC) representing the community of all member states [as collective trustor/settlor];
(b) the host state of a site [as trustee] incurs fiduciary duties to protect and conserve the site so dedicated for the benefit of present and future generations of "all the peoples of the world" [as beneficiaries], and to report to the trustor [and the co-trustees] through the WHC on the conservation status of the site (so-called 'active monitoring'); and
(c) the beneficiaries, represented by civil society organizations, may invoke the terms of the trust to hold the host/trustee state accountable for non-compliance with the terms of the trust, either through their national courts,[13] or through the WHC by requesting the down-listing of a site as "world heritage in danger", or eventual de-listing ('reactive monitoring'; UNESCO 2012, s. 169–174; Litton 2011, p. 234).[14] In view of its wide transnational media attention in particular,

[12]Curiously though, the term was retained in article 15(2) of the Convention for the 'world heritage fund' (a "trust fund" in the English text, officially translated into French as "*fonds de dépot*" and into Spanish as "*fondo fiduciario*").

[13]E.g., see the decision of the Federal Court of Australia in *Friends of Hinchinbrook Society v. Minister for Environment* 1997, confirming an NGO's standing to challenge governmental decisions concerning the Great Barrier Reef world heritage site, and several other world heritage cases brought by NGOs in Australian courts (Boer and Wiffen 2006); see also the South African High Court decision in *Hout Bay Residents' Association et al. v. Entillini Concession Ltd* 2012.

[14]A critical IUCN report thus triggered action by the World Heritage Committee in the case of Australia's Kakadu National Park (Morgera 2009, p. 228). A similar down-listing scheme for endangered sites, albeit based on unilateral governmental site nominations, has been developed under the Ramsar Convention 1971, through its 'Montreux Record' created by decision VI.1/1996 of the Conference of the Parties.

the WHC down-listing practice thus evolved into an effective participatory instrument to induce compliance with the trusteeship regime (Redgwell 2002; Battini 2011; Francioni and Gordley 2013).

One generation later, the Assembly of the Food and Agriculture Organization of the United Nations (FAO) adopted its *Plant Genetic Resources Treaty* (ITPGR 2001; Raustiala and Victor 2004; Moore and Tymowski 2005), which in fact confirmed and consolidated international trusteeship status for 12 of the world's major *ex-situ* germplasm collections under the auspices of the Consultative Committee on Agricultural Research (CGIAR; see Siebeck and Barton 1992; Moore and Frison 2011):

(a) the germplasm material listed in Annex I of the treaty (including wild predecessors of 35 cultivated food crop genera and 29 forage species) is designated/dedicated as the *corpus* of the trust, pursuant to a model "in-trust agreement" under which the host States and institutions [as trustees] agree to "hold the designated germplasm in trust for the benefit of the international community, in particular the developing countries" (see Gotor et al. 2010);
(b) transnational access under the multilateral system is governed by a standardized materials transfer agreement adopted in 2006, which also addresses benefit-sharing issues—in somewhat uneasy coexistence with the *Biodiversity Convention* 1992 (Lochen 2007, pp. 228–229) and its *Nagoya Protocol* 2010 (Moore and Williams 2011; Chiarolla et al. 2012; Biber-Klemm et al. 2013, p. 219); and
(c) compliance is monitored by a Compliance Committee reporting to the treaty's Governing Body (ITPGR 2011). Critics have pointed out, however, that this accountability mechanism provides as yet only very limited opportunities for participation by civil society beneficiaries, represented predominantly by business stakeholders (Mooney 2011, pp. 145–148).

What is significant here is that both the UNESCO World Heritage regime and the FAO Plant Genes regime currently apply the 'global commons' label solely to resources situated *within* the territorial jurisdiction of States.[15] In fact, international trusteeship is not only quite compatible with customary territorial sovereignty, but its operation in practice necessarily relies on States exercising sovereign powers, albeit on behalf of the global community, through a kind of 'role-splitting' ("*dé doublement fonctionnel*"; Scelle 1932, pp. 54–56, 217; Scelle 1956; Cassese 1990). In this regard, there has indeed been something of a paradigm change in the perception of the sovereignty dilemma by environmentalists: The early literature of international environmental law and governance had started out from a radical iconoclastic *critique* of the 'formidable defensive concept' of permanent sovereignty of States over natural resources (Allott 1989), suspected to lurk at the roots

[15]Halewood et al. (2012); see, however, recent proposals to apply world heritage criteria also to resources in high sea areas (Abdulla et al. 2013, pp. 46–47; Freestone et al. 2016).

of many global environmental problems (Falk 1971, p. 222; [16] Sprout and Sprout 1971, p. 406; Caldwell 1973, p. 200); and from high hopes for a 'fading away', 'erosion', or 'perforation' of territorial sovereignty, as the preferred solution to those problems (Mayer-Tasch 1985; Van der Lugt 2000; Karkkainen 2004). As the subsequent evolution of global and regional lawmaking in this field demonstrated, however, 'proprietary' sovereign rights can effectively be limited and balanced by overriding 'fiduciary' or 'custodial' community norms, provided those are backed up by the necessary procedural mechanisms to hold States accountable as trustees (Sand 2004; Scholtz 2008).

Admittedly, my empirical examples are still fragmentary, and a far cry from the grand design of the less patient advocates of 'earth governance' among us. Basic questions remain, in particular, as to the most appropriate representation of an international trust's beneficiaries; i.e., present and future civil society (Bothe 2006, p. 555; Bosselmann 2015, pp. 252–257).[17] Yet, the idea of public trusteeship for the commons may rightfully be counted among Nino Cassese's 'realistic utopia' of international law (Cassese 2012; Francioni 2012, p. 443; Allott 2014), or rather the '*eutopia*' envisioned by Allott (2016).[18] And even if it were a mere 'mobilizing myth' invented by environmental lawyers (as suggested by René Dupuy 1985, p. 504; using a term coined by Georges Sorel 1908, p. 141),—a *fata morgana*, a mirage—we should perhaps keep in mind Leszek Kolakowski's alternative image of a mirage,

> which makes beautiful lands arise before the members of a caravan and thus increases their efforts to the point where, in spite of all their sufferings, they reach the next tiny waterhole. Had such tempting mirages not appeared, the exhausted caravan would inevitably have perished in the sandstorm, bereft of hope (Kolakowski 1988, p. 32).

References

Aarhus Convention (1998) Convention on Access to Information, Public Participation in Decision-Making and Access to Justice in Environmental Matters, 25 June 1998. 2161 U.N.T.S. 447

Abdulla AA, Obur D, Bertzky B, Shi Y (2013) Marine natural heritage and the World Heritage List. International Union for Conservation of Nature, Gland

[16]But see Falk (1995), p. 11: "I now believe that this earlier analysis was badly mistaken in several key respects."

[17]E.g., a 'global commons trusteeship commission' as proposed by Cleveland (1993); an 'ombudsman' or 'environmental high commissioner' as proposed by Orrego Vicuña and Sohn (1997), pp. 288, 341; or global commons 'guardians', as suggested by Stone (1993), pp. 39–43, and Sands (1997), p. 83.

[18]The vision of a *eutopia futuris* goes back to Scottish town planner Patrick Geddes, in his lectures to the British Sociological Society (Geddes 1905), to describe a 'good place' of the future, a place that can be achieved through local and international cooperation, and adoption of sustainable technologies; as distinct from *utopia*, as an ideal place impossible to achieve. It also appears on one of the stained-glass windows of his Outlook Tower in Edinburgh.

Allott P (1989) International law and international revolution: reconceiving the world. Josephine Onoh Memorial Lecture, University of Hull Press, Hull; reprinted in Freestone D, Subedi S, Davidson S (eds) Contemporary issues in international law (2002). Kluwer Law International, The Hague, pp 77–98

Allott P (2001) Eunomia: new order for a new world, 2nd edn. Oxford University Press, Oxford

Allott P (2002) The health of nations: society and law beyond the state. Cambridge University Press, Cambridge

Allott P (2014) The idealist's dilemma: re-imagining international society. Eur J Int Law Blog: EJIL Talk!, 9 June 2014

Allott P (2016) Eutopia: new philosophy and new law for a troubled world. Edward Elgar, Cheltenham

Ardrey R (1966) The territorial imperative: a personal inquiry into the animal origins of property and nations. Dell Publishing, New York

Baber WF, Bartlett RV (2016) Democratic accountability in the anthropocene. In: Pattberg P, Zelli F (eds) Environmental politics and governance in the anthropocene: institutions and legitimacy in a complex world. Routledge, Abingdon, pp 167–183

Barritt E (2015) Standing up for British lungs: effective judicial enforcement in environmental law – R (on the application of ClientEarth) v. Secretary of state for the environment, food and rural affairs. Rev Eur Comp Int Enviro Law 24:368–372

Batisse M, Bolla G (2003) L'invention du 'patrimoine mondial'. In: UNESCO action as seen by protagonists and witnesses, History Paper No. 2. Association of Former UNESCO Staff Members, Paris

Battini S (2011) The procedural side of legal globalization: the case of the World Heritage Convention. Int J Const Law 9:340–368

Benvenisti E (2013) Sovereigns as trustees of humanity: on the accountability of states to foreign stakeholders. Am J Int Law 107:295–333

Biber-Klemm S et al (2013) Governance options for *ex-situ* collections in academic research. In: Oberthür S, Rosendal GK (eds) Global governance of genetic resources: access and benefit sharing after the Nagoya Protocol. Routledge, Abingdon, pp 213–230

Biodiversity Convention (1992) Convention on Biological Diversity, 5 June 1992. 1760 U.N.T.S. 79

Boer B, Wiffen G (2006) Heritage law in Australia. Oxford University Press, Oxford

Bosselmann K (2015) Earth governance: trusteeship of the global commons. Edward Elgar, Cheltenham

Bothe M (2006) Whose environment? Concepts of commonality in international environmental law. In: Winter G (ed) Multilevel governance of global environmental change: perspectives from science, sociology and the law. Cambridge University Press, Cambridge, pp 539–558

Brown Weiss E (1989) In fairness to future generations: international law, common patrimony, and intergenerational equity. United Nations University and Transnational Publishers, Dobbs Ferry, NY

Brunnée J (2005) Legal accountability through the lens of the law of state responsibility. Neth Yearb Int Law 36:3–38

Brunnée J (2007) Common areas, common heritage, and common concern. In: Bodansky D, Brunnée J, Hey E (eds) Oxford handbook of international environmental law. Oxford University Press, Oxford, pp 550–573

Buck SJ (1998) The global commons: an introduction. Island Press, Washington, DC

Caldwell LK (1973) Concepts in development of international environmental policies. Natural Resour J 13:190–212

Candioti E (2014) Statement at the 3212th Meeting of the International Law Commission (30 June 2014). Summary Record: UN Doc A/CN.4/SR.3212

Capra F, Mattei U (2015) The ecology of law: toward a legal system in tune with nature and community. Berrett-Koehler, San Francisco

Carrington D (2016) High Court rules UK government plans to tackle air pollution are illegal. The Guardian, London, 2 Nov 2016

Cassese A (1990) Remarks on Scelle's theory of 'role splitting' (*dédoublement fonctionnel*) in international law. Eur J Int Law 1:210–231

Cassese A (ed) (2012) Realizing utopia: the future of international law. Oxford University Press, Oxford

Chiarolla C, Louafi S, Schloen M (2012) An analysis of the relationship between the Nagoya protocol and instruments related to genetic resources for food and agriculture and farmers' rights. In: Morgera E, Buck M, Tsioumani E (eds) The Nagoya protocol on access and benefitsharing in perspective: implications for international law and implementation challenges. Martinus Nijhoff, Leiden, pp 83–122

Cleveland H (1990) The global commons: policy for the planet. University Press of America, Lanham

Cleveland H (1993) The global commons: a global commons trusteeship commission is needed to guide our use of the oceans, Antarctica, the atmosphere, and outer space. Futurist 27(3):9–13

ClientEarth v. Department for the Environment, Food and Rural Affairs (2015) [2015] UKSC 28, 29 April 2015

Coghill K, Sampford C (2012) In: Smith T (ed) Fiduciary duty and the atmospheric trust. Routledge, Abingdon

Deutsche Umwelthilfe, Verkehrsclub Deutschland v. Ministry of Environment and Consumer Protection (2016), Munich *Verwaltungsgericht* (Administrative Tribunal, First Chamber), 29 June 2016. Cases No. M1V 15.5203 (confirming an earlier judgment of 9 Oct 2012), and No. M1K 15.5714

Drion H (1954) Limitation of liabilities in international air law. Springer, Dordrecht

Dubnick MJ (2014) Accountability as a cultural keyword. In: Bovens M, Goodin RE, Schillemans T (eds) Oxford handbook of public accountability. Oxford University Press, Oxford, pp 23–38

Dupuy RJ (1985) Conclusions. In: Dupuy RJ (ed) L'avenir du droit international de l'environnement (The future of the international law of the environment). Martinus Nijhoff, Dordrecht, pp 497–514

Durner W (2001) Common goods: Statusprinzipien von Umweltgütern im Völkerrecht. Nomos, Baden-Baden

Ebbesson J, Gaugitsch H, Jendroska J, Stec S (2014) The Aarhus Convention: an implementation guide. United Nations Economic Commission for Europe, Geneva

Epiney A (2006) The role of NGOs in the process of ensuring compliance with MEAs. In: Beyerlin U, Stoll PT, Wolfrum R (eds) Ensuring compliance with multilateral environmental agreements: a dialogue between practitioners and academia. Martinus Nijhoff, Leiden, pp 319–352

Everard M, Pontin B, Appleby T, Staddon C, Hayes ET, Barnes J, Longhurst J (2013) Air as a common good. Environ Sci Policy 33:354–368

Falk RA (1971) This endangered planet: prospects and proposals for human survival. Random House, New York

Falk RA (1995) Environmental protection in an era of globalization. Yearb Int Environ Law 6:3–25

Faure M, Peeters M (2011) Climate change liability. Edward Elgar, Cheltenham

Fitzmaurice M (1996) Liability for environmental damage caused to the global commons. Rev Eur Comp Int Environ Law 5:305–311

Francioni F (2012) Realism, utopia, and the future of international environmental law. In: Cassese A (ed) Realizing utopia: the future of international law. Oxford University Press, Oxford, pp 442–458

Francioni F, Gordley J (2013) Enforcing international cultural heritage law. Oxford University Press, Oxford

Fratcher WF (1973) Trust. Int Encycl Comp Law 6(11). Mohr Siebeck, Tübingen 1973, pp 84–141

Freestone D, Laffoley D, Douvere F, Badman T (eds) (2016) World heritage in the seas: an idea whose time has come. UNESCO/IUCN, Paris

Friends of Hinchinbrook Society v. Minister for Environment (1997) Federal Court of Australia, Aug 1997. Aust Law Rep 142:632 and 147:608

Gabcikovo-Nagymaros case (1997) International Court of Justice, 25 Sept 1997. Separate opinion of Judge Christopher Weeramantry; reprinted in: International Legal Materials 37:204

Geddes P (1905) Civics: as applied sociology. Sociological Papers 1:104–144; reprinted in: Meller HE (ed) The ideal city. Leicester University Press, Leicester, pp 75–122

Gotor E, Caracciolo F, Watts J (2010) The perceived impact of the in-trust agreements on CGIAR germplasm availability: an assessment of bioversity international's institutional activities. World Dev 38:1486–1493

Gutzwiller M (1947) In memoriam Karl Neumeyer. Annuaire de l'Institut de Droit International 317–322

Halewood M, López Noriega I, Louafi S (eds) (2012) Crop genetic resources as a global commons: challenges in international law and governance. Routledge, Abingdon

Halfmann J (2012) Die Atmosphäre als *Global Common*: wissenschaftliche und politische Adressierung. In: Morisse-Schilbach M (ed) Wissen, Wissenschaft und *Global Commons*. Nomos, Baden-Baden, pp 133–152

Hansmann H, Mattei U (1998) The functions of trust law: a comparative legal and economic analysis. N Y Univ Law Rev 73:434–479

Hardin G (1968) The tragedy of the commons. Science 162:1243–1248

Harrison J, Matson P (2001) The atmospheric commons. In: Burger J, Ostrom E, Norgaard R, Policansky D, Goldstein BD (eds) Protecting the commons: a framework for resource management in the Americas. Island Press, Washington, pp 219–239

Hout Bay Residents' Association et al. v. Entillini Concession Ltd (2012) South African Western Cape High Court, 6 June 2012. Case No. 7648/12

ILC (2001) Report of the International Law Commission on its 53rd Session. UN Doc A/56/10

ILC (2013) Report of the International Law Commission on its 65th Session. UN Doc A/68/10

ILC (2015) Report of the International Law Commission on its 67th Session. UN Doc A/70/10

ITPGR (2001) International Treaty on Plant Genetic Resources for Food and Agriculture, 2 Nov 2001. 2400 U.N.T.S. 303

ITPGR (2011) Resolution 2/2011 on Procedures and Oparationl Mechanisms to Promote Compliance and Address Issues of Non-Compliance (Bali, 18 March 2011). Summary in Earth Negotiations Bulletin 9:550 (21 March 2011), pp 4–6

ITU (1992) (International Telecommunication Union) Constitution. Geneva, 2015 edn

IUCN (2015) Draft international covenant on environment and development, updated 5th edn. International Union for Conservation of Nature, Gland

Jacobs R (2004) Mandat und Treuhand im Völkerrecht. Universitätsdrucke, Göttingen

Jonas H (1984a) Das Prinzip Verantwortung: Versuch einer Ethik für die technologische Zivilisation, 4th edn. Suhrkamp, Frankfurt

Jonas H (1984b) The imperative of responsibility: in search of an ethics for the technological age. University of Chicago Press, Chicago

Joyner CC (2001) Global commons: the oceans, Antarctica, the atmosphere and outer space. In: Simmons PJ, de Jonge OC (eds) Managing global issues: lessons learned. Carnegie Foundation, Washington

Kain et al. v. Department of Environmental Protection (2016) Massachusetts Supreme Judicial Court, 17 May 2016. Docket No. SJC-11961

Karkkainen BC (2004) Post-sovereign environmental governance. Glob Environ Polit 4:72–96

Keohane RO (2003) Global governance and democratic accountability. In: Held D, Koenig-Archibugi M (eds) Taming globalization: frontiers of governance. Polity Press, Oxford, pp 130–159

Keohane RO, Victor DG (2011) The regime complex for climate change. Perspect Polit 9:7–23

Khan DE (2012) Territory and boundaries. In: Fassbender B, Peters A (eds) Oxford handbook of the history of international law. Oxford University Press, Oxford, pp 225–249
Kish J (1973) The law of international spaces. Sijthoff, Leiden
Kiss AC (1982) La notion de patrimoine commun de l'humanité. Hague Academy Int Law Collect Courses 175:145–151
Kiss AC, Shelton D (2007) Guide to international environmental law. Martinus Nijhoff, Leiden
Kolakowski L (1988) Der Mensch ohne Alternative. Piper Verlag, Munich; as translated in Hirschman AO (1995) Development projects observed, rev. edn. Brookings Institution Press, Washington
Koskenniemi M (2005) International law as therapy: reading The Health of Nations. Eur J Int Law 16:329–341
Kramarz T, Park S (2016) Accountability in global environmental governance: a meaningful tool for action? Glob Environ Polit 16:1–21
Lee RJ (2012) Law and regulation of commercial mining of minerals in outer space. Springer, Dordrecht
Lewis JS (2014) Asteroid mining 101: wealth for the new space economy. Deep Space Industries, Moffett
Litton S (2011) The world heritage 'in danger' listing as a taking. N Y Univ J Int Law Polit 44:219–265
Lochen T (2007) Die völkerrechtlichen Regelungen über den Zugang zu genetischen Ressourcen. Mohr Siebeck, Tübingen
Lode B, Schönberger P, Toussaint P (2016) Clean air for all by 2013? Air quality in the 2030 agenda and in international law. Rev Eur Comp Int Law 25:27–38
Lord R, Goldberg S, Rajamani L, Brunnée J (eds) (2011) Climate change liability: transnational law and practice. Cambridge University Press, Cambridge
Lyall F (2011) International telecommunications: the international telecommunication union and the universal postal union. Routledge, Abingdon
Lyall F, Larsen PB (2009) Space law: a treatise. Routledge, Abingdon
MacWhorter K (2016) Sustainable mining: incentivizing asteroid mining in the name of environmentalism. William Mary Environ Law Policy Rev 40:645–676
Marx K (1865) Das Kapital, vol 3 ch 46. In: Engels F (ed) Ökonomische Manuskripte 1863-1867, reprinted in Marx-Engels-Gesamtausgabe, Berlin 1992, part II vol 4
Mattei U (2015) Il benicomunismo e i suoi nemici. Einaudi, Turin
Matz N (2005) Civilization and the mandate system under the League of Nations as origins of trusteeship. Max Planck Yearb United Nations Law 9:47–96
Mayer-Tasch PC (1985) Internationale Umweltpolitik als Herausforderung für die Nationalstaatlichkeit. Das Parlament: Aus Politik und Zeitgeschichte 20:3–13
Meyer RL (1976) *Travaux préparatoires* for the UNESCO World Heritage Convention. Earth Law J 2:45–81
Milun K (2011) The political uncommons: the cross-cultural logic of the global commons. Ashgate, Farnham
Minca C, Rowan R (2015) The question of space in Carl Schmitt. Prog Hum Geogr 39:268–289
Monro CH (1904) The digest of Justinian, vol 1. Cambridge University Press, Cambridge
Mooney P (2011) International non-governmental organizations: the hundred-year (or so) seed war – seeds, sovereignty and civil society – a historical perspective on the evolution of 'the law of the seed'. In: Frison C, López F, Esquinas-Alcázar JT (eds) Plant genetic resources and food security: stakeholder perspectives on the international treaty on plant genetic resources for food and agriculture. Earthscan, London, pp 135–148
Moore G, Frison E (2011) International research centres: the Consultative Committee on International Agricultural Research and the international treaty. In: Frison C, López F, Esquinas-Alcázar JT (eds) Plant genetic resources and food security: stakeholder perspectives on the international treaty on plant genetic resources for food and agriculture. Earthscan, London, pp 149–162

Moore G, Tymowski W (2005) Explanatory guide to the International Treaty on Plant Genetic Resources for Food and Agriculture. IUCN Environmental Policy and Law Paper No. 57. IUCN, Cambridge

Moore G, Williams KA (2011) Legal issues in plant germplasm collecting. In: Guarino L, Ramanath Rao V, Reid R (eds) Collecting plant diversity: technical guidelines. Bioversity International, Rome

Morgenthau HJ (1941) Karl Neumeyer obituary. Am J Int Law 35:672

Morgera E (2009) Corporate accountability in international environmental law. Oxford University Press, Oxford

Mulgan R (2000) Accountability: an ever expanding concept? Public Adm 78:555–573

Murase S (2011) Protection of the atmosphere. Report of the International Law Commission on its 63rd Session, UN Doc A/66/10, Annex B, pp 315–329

Murase S (2012) Protection of the atmosphere and international law: rationale for codification and progressive development. Sophia Law Rev 55:1–58

Nagoya Protocol (2010) Protocol on Access to Genetic Resources and the Fair and Equitable Sharing of Benefits Arising from their Utilization, 29 Oct 2010. UN Doc UNEP/CBD/COP/DEC/X/1

Najam A, Halle M (2010) Global environmental governance: the challenge of accountability. Sustainable Dev Insights 5:1–8

Neumeyer KA (1922) Internationales Verwaltungsrecht, vol 2. Schweitzer, Munich

Obama B, Chen J, Perelli T, Hay L, Cohen J (eds) (1991) Developments in the law: international environmental law. Harv Law Rev 104:1484–1639

Orrego Vicuña F, Sohn L (1997) Responsibility and liability under international law for environmental damage. Annuaire de l'Institut de Droit International I:288, I:341

Ostrom E (1990) Governing the commons: the evolution of institutions for collective action. Cambridge University Press, Cambridge

Ostrom E (1999) Die Verfassung der Allmende: jenseits von Staat und Macht. Mohr Siebeck, Tübingen

Ostrom E (2011) Was mehr wird, wenn wir teilen: vom gesellschaftlichen Wert der Gemeingüter. Oekom, Munich

Peeters M (2016) *Urgenda Foundation and 886 Individuals v. The State of the Netherlands*: The dilemma of more ambitious greenhouse gas reduction action by EU member states. Rev Eur Comp Int Environ Law 25:123–129

Peter C (2015) Statement at the 3247th Meeting of the International Law Commission (8 June 2015). Summary Record: UN Doc A/CN.4/SR.3247

Pitea C (2005) NGOs in non-compliance mechanisms under multilateral environmental agreements: from tolerance to recognition. In: Treves T (ed) Civil society, international courts and compliance bodies. T.M.C. Asser Press, The Hague, pp 205–224

Plakokefalos I (2013) International Law Commission and the topic "protection of the atmosphere": anything new on the table? SHARES blog (Amsterdam, 1 Nov 2013), available at http://www.sharesproject.nl

Rached DH (2016) The concept(s) of accountability: form in search of substance. Leiden J Int Law 29:317–342

Ramsar Convention (1971) Convention on wetlands of international importance especially as waterfowl habitat UNTS 996:245

Raustiala K, Victor DG (2004) The regime complex for plant genetic resources. Int Org 58:277–309

Razzaque J (2001) Case law analysis: application of public trust doctrine in Indian environmental cases. J Environ Law 13:221–234

Redgwell C (1999) Intergenerational trusts and environmental protection. Manchester University Press, Manchester

Redgwell C (2002) The international law of public participation: protected areas, endangered species, and biological diversity. In: Zillman DM, Lucas A, Pring G (eds) Human rights in

natural resource development of mineral and energy resources. Oxford University Press, Oxford
Redgwell C (2007) Protecting natural heritage and its transmission to future generations. In: Yussuf AA (ed) Standard-setting in UNESCO: normative action in education, science and culture, vol 1. Martinus Nijhoff, Leiden
Richard V (2011) L'*accountability* comme alternative à la responsabilité? Réflexions en droit international de l'environnement. In: Vergès E (ed) Droit, sciences et techniques: quelles responsabilités? LexisNexis, Paris, pp 523–541
Rodotà S (2013), Il terribile diritto: studi sulla proprietà privata e sui beni comuni. Mulino, Bologna
Rubin A (1993) Comment. In: Delbrück J (ed) The future of international law enforcement: new scenarios – newlaw? Duncker & Humblot, Berlin, pp 171–174
Ryan PS (2004) Application of the public trust doctrine and principles of natural resource management to electromagnetic spectrum. Michigan Telecommun Technol Law Rev 10:285–372
Sand PH (2004) Sovereignty bounded: public trusteeship for common pool resources? Glob Environ Polit 4(1):47–71
Sand PH (2012) The evolution of transnational environmental law: four cases in historical perspective. Transnational Environ Law 1:183–198
Sand PH (ed) (2015) The history and origin of international environmental law. Edward Elgar, Cheltenham
Sand PH, Wiener JB (2016) Towards a new international law of the atmosphere? Goettingen J Int Law 7(2):195–223
Sanders TC (1903) The institutes of Justinian, 4th edn. Longmans Green, London
Sands PJ (1989) The environment, community and international law. Harv Int Law J 30:393–420
Sands PJ (1997) Protecting future generations: precedents and practicalities. In: Agius E, Busuttil S (eds) Future generations and international law. Routledge, Abingdon, pp 83–91
Scelle G (1932) Précis de droit des gens: principes et systématique, vol 1. Sirey, Paris
Scelle G (1956) Le phénomène du dédoublement fonctionnel. In: Schätzel W, Schlochauer HJ (eds) Rechtsfragen der internationalen Organisation: Festschrift für Hans Wehberg. Frankfurt, pp 324-342
Scelle G (1958) Obsession du territoire. In: Van Asbeck FM (ed) Symbolae Verzijl. T.M.C. Asser Press, The Hague, pp 347–361
Schmitt C (1997) Der Nomos der Erde im Völkerrecht des *Ius Publicum Europaeum*, 5th edn. Duncker and Humblot, Berlin; English transl. Ulmen GL (2006) The nomos of the earth in the international law of the *ius publicum europaeum*. Telos Press, New York
Scholtz W (2008) Custodial sovereignty: reconciling sovereignty and global environmental challenges amongst the vestiges of colonialism. Neth Int Law Rev 55:323–344
Schrijver N (2016) Managing the global commons: common good or common sink? Third World Q 37:1252–1267
Scobbie I (2005) Slouching towards the holy city: some weeds for Philip Allott. Eur J Int Law 16:299–313
Siebeck WE, Barton JH (1992) The implications of applying the legal concept of trust to germplasm collections at CGIAR research centres. Diversity 8(3):29–35
Sinclair A (1995) The chameleon of accountability: forms and discourses. Account Org Soc 20:219–237
Sorel G (1908) Réflexions sur la violence. Rivière, Paris
Soroos MS (1997) The endangered atmosphere: preserving a global commons. University of South Carolina Press, Columbia
Soroos MS (1998) The thin blue line: preserving the atmosphere as a global commons. Environment 40(2):6–13
Sprout HH, Sprout M (1971) Towards a politics of the Planet Earth. Van Nostrand Reinhold, New York

Stone CD (1993) Defending the global commons. In: Sands PJ (ed) Greening international law. Routledge, Abingdon, pp 34–49
Tams CJ (2011) Individual states as guardians of community interests. In: Fastenrath F, Geiger R, Khan DE, Paulus A, von Schorlemer S, Vedder C (eds) From bilateralism to community interest: Essays in honour of Judge Bruno Simma. Oxford University Press, Oxford, pp 379–205
Taylor P, Stroud L (2013) Common heritage of mankind: a bibliography of legal writings. Malta, Fondation de Malte
Tomuschat C (1993) The global commons. Yearb Int Law Commission 2(1):245–247
Train RE (1972) A world heritage trust. In: Gillette ER (ed) Action for wilderness. Sierra Club, Washington, pp 172–176
Treves T, Pineschi L, Pitea C, Ragni C, Romanin Jacur F (eds) (2009) Non-compliance procedures and mechanisms, and the effectiveness of international environmental agreements. T.M.C. Asser Press, The Hague
U.S. Space Resource Exploration and Utilization Act (2015) Public Law 114-90 of 25 Nov 2015, 129 Stat. 721,51 U.S. Code s. 51303
UNESCO (1972) *Convention for the Protection of the World Cultural and Natural Heritage*, 23 Nov 1972. 1037 U.N.T.S. 151
UNESCO (2012) *Operational Guidelines for the Implementation of the World Heritage Convention*, July 2012. UNESCO Doc WHC.12/01
UNFCCC (1992) *United Nations Framework Convention on Climate Change*, 9 May 1992. 1771 U.N.T.S. 107
UNGA 2011 (*United Nations General Assembly*, Sixth Committee) Summary Records of the 20th Meeting, 23 Nov 2011. UN Docs A/C.6/66/SR.19 (para. 8) and A/C.6/66/SR.20 (paras. 15 and 48)
Urgenda et al. v. Ministry of Infrastructure and the Environment (2015) The Hague District Court, 24 June 2015.Case No. C/09/456689/HA-ZA 13-1396, English transl. ECLI:NL: RBDHA:2015:7196
Van der Lugt CT (2000) State sovereignty or ecological sovereignty: a study of acid rain within the European Union. Nomos, Baden-Baden
Van der Schyff E (2013) Stewardship doctrines of public trust: has the eagle of public trust landed on South African soil? South Afr Law J 130:369–389
Van Putten M (2008) Policing the banks: accountability mechanisms for the financial sector. McGill University Press, Montreal
Vogel K (1970) In memoriam Karl Neumeyer. Archiv des öffentlichen Rechts 95:138–140
Vogel K (2001) Karl Neumeyer: In den Tod getrieben. In: Landau P, Nehlsen H (eds) Große jüdische Gelehrte an der Münchener juristischen Fakultät. Aktiv, Ebelsbach, pp 97–111
Vogler J (2001) Future directions: the atmosphere as a global commons. Atmos Environ 35:2427–2428
Vogler J (2012) Global commons revisited. Glob Policy 3:61–71
Von Breitenbuch H (2013) Karl Neumeyer: Leben und Werk, 1869–1941. Lang, Frankfurt
Von Schorlemer S (2012) Telecommunications: international regulation. In: Wolfrum R (ed) Max Planck Encyclopedia of Public International Law 9:818–841
Wall D (2014) The commons in history: culture, conflict, and ecology. MIT Press, Cambridge
Waters DWM (1995) The institution of the trust in civil and commercial law. Hague Acad Int Law Collect Courses 252:113–454
Wehberg H (1941) Karl Neumeyer obituary. Die Friedenswarte 41:256–260
Westra L (1994) An environmental proposal for ethics: the principle of integrity. Rowman Littlefield, Lanham
Westra L (2016) Ecological integrity and global commons: science, ethics and the law. Routledge, Abingdon
Wheatcroft A (1996) The Habsburgs: embodying empire. Penguin, London
Wolfrum R (1984) Die Internationalisierung staatsfreier Räume. Springer, Heidelberg

Wood MC (2009) Advancing the sovereign trust of government to safeguard the environment for present and future generations: instilling a fiduciary obligation in governance. Environ Law 39:91–139

Wood MC (2013) Nature's trust: environmental law for a new ecological age. Cambridge University Press, Cambridge

Wood MC, Woodward CW IV (2016) Atmospheric trust litigation and the constitutional right to a healthy climate system: judicial recognition at last. Wash J Environ Law Policy 62:633–683

Wustlich G (2003) Die Atmosphäre als globales Umweltgut: Rechtsfragen ihrer Bewirtschaftung im Wechselspiel von Völker-, Gemeinschafts- und nationalem Recht. Duncker Humblot, Berlin

Integrity at Risk: Potentials and Dangers of Synthetic Biology and How to Govern with Integrity

Franz-Theo Gottwald

1 The Start of a Revolution

Recently a renowned researcher gave a presentation on the new biotechnologies at Munich Technical University (TUM). Angelika Schnieke's academic resume is impressive! Professor Schnieke (b. 1956) has many years of experience in the manufacture of pharmaceutical products in the milk of livestock. Her Chair at TUM conducts research into the application of biotechnology in large animals to address problems in agricultural sciences in general and biomedicines in particular. Her research interests are in the areas of animal stem cells, the genetic modification of mammals for regenerative medicine and the production of animal models of human diseases.

Professor Schnieke studied bioengineering in Hamburg and obtained her Ph.D. in Medicine from the University of Edinburgh. After her studies she worked at a number of German and international research institutes (Heinrich-Pette Institute in Hamburg; Whitehead Institute at MIT in Boston, Ludwig Institute in Bern and Colorado State University). Before being appointed professor at TUM she worked at biotechnology company PPL Therapeutics, Edinburgh, first as Head of Molecular Biology and later as Assistant Director of Research. (WZW 2016).

I cite Professor Schnieke's academic resume because it is symptomatic for many experts who are in the process of revolutionizing biology. Some of them work on plants, some focus on animals (like Prof. Schnieke). Gene-editing is the central focus of Professor Schnieke's research, promising genetic revolution in the livestock sector.

F.-T. Gottwald (✉)
Albrecht Daniel Thaer-Institute of Agricultural and Horticultural Sciences, Humboldt University Berlin, c/o Schweisfurth Foundation, Munich, Germany
e-mail: Post@Schweisfurth.de

L. Westra et al. (eds.), *The Role of Integrity in the Governance of the Commons*,
DOI 10.1007/978-3-319-54392-5_2

Starting with the domestication of animals more than 10,000 years ago, the genetic makeup of livestock has been specifically tampered with, i.e. one generation after the other, mostly on farms and based on experience and exchange between farmers. But in addition to conventional breeding during the past 30 years genetic engineering has enabled introduction of desired traits in large animals. Even though there are several examples illustrating the benefits for agriculture, however, it has been in the biomedical sector where advanced research and technological developments in livestock have taken place over the past years. Dolly, the cloned sheep was the first breakthrough 20 years ago. For the first time the genome could be altered in such a way as to influence individual traits. Given that this is a very complicated technical process, there are only a few labs around the world where this kind of genome modification has been achieved successfully. Nowadays there is gene-editing: it is simple, workable and efficient. It works for plants, for insects, fish and all the way up to mammals. Sequencing allows reading of the genome; gene-editing allows for alteration or editing, much like in a computer program. Mankind nowadays is capable of exchanging individual bases or copy and correct natural mutations. Depending on the eyes of the beholder, an advantage or disadvantage of gene-editing is the fact that, depending on the application chosen, an edited animal is indistinguishable from a natural mutation or variation. In some countries—like the US—this new technology is applied in agriculture with great optimism and innovative zeal, and newspapers use headlines like: "Genetic Engineering for the Cowshed"; "Food and Genome-Editing: Genetic Engineering Revolution in the Field and on the Table?" The time has come to address this genetic engineering revolution in livestock.

There is no doubt that bioethicists and researchers in agricultural ethics have to address this topic of ecological integrity in science and technology, given that culturally developed notions on the dignity of living beings and the need for safeguarding individual lives of animals and plants would be radically challenged.

2 Synthetic Biology: Potentials and Dangers

In the new broader scientific framework of bioeconomy (with gene-editing being one component) synthetic biology undoubtedly is most challenging to the integrity of life forms. Synthetic Biology—or Extreme Genetic Technology as labeled by others—promises more than innovations: *How Synthetic Biology Will Reinvent Nature and Ourselves* is the title of the programmatic book by George Church, together with Craig Venter the most popular and effective prophet of synthetic biology. This is quite a tall order. Synthetic biology goes a major step further than "classic genetic technologies"; it not only aims at genetically altering life forms, but rather its aim is a complete rearrangement or new construction of life. The idea of basically being able to take any bacterium, microbe or algae, build a kind of miniature processing facility, which, fed with practically any sort of biomass will produce whatever is desired (e.g. fuel, plastics, vanilla flavoring etc.) goes way beyond classical methods of genetic engineering. This kind of vision comprises a totally novel production process, and consequently a new type of economics—rendering nature into a kind of magic grab bag and man the ruler over nature and all natural processes. The term "natural" thus becomes meaningless as well as the concept of integrity.

At the moment synthetic biology has been used for the large-scale production of next-generation biofuels. This has motivated large carbon manufacturers as well as numerous international energy and chemical industry companies like Shell, Exxon, BP, Chevron, Total, Petrobras, BASF, Dow and DuPont during the past years to invest billions into small and mid-sized start-ups from Silicon Valley. Due to a large number of technical problems (predominantly with regard to increased productivity), the industry increasingly turned toward other products during the past years, i.e. "high-value" and "low-volume" products such as flavorings and additives for the food and the cosmetics industries. Whereas until recently markets predominantly favored mass chemicals, bioplastics and biofuels, the past years saw the introduction of synthetic vanilla flavoring by Evolva, Artemisinin (an anti-malarial agent) by Amyris or lauric acid (used in soap making) by Solazyme. Others like synthetic-biology rose oil, Stevia, sandal wood, saffron and lactic acid are close to introduction (as of January 2015).

It goes without saying that defense and health ministries have been showing much interest in synthetic biology with regard to the introduction of new medical and military miracle weapons designed as defensive weapons against enemy attacks using the same methods.

The industry is very dynamic. But resistance is growing. One example: The Belgian cleanser and detergent manufacturer Ecover has made the announcement that palm seed oil will be substituted with algae oil in some of its products. Palm seed oil is pressed from palm kernels, whereas palm oil from tree fruit. Detergent manufacturers like Ecover argue that ecological gains result from a reduction in harmful palm oil production. It can be argued that cultivation of essential biomass (sugar) to feed the algae will offset the desired reduction in land under cultivation. But assuming that a positive balance in cultivated areas will be possible—the oil is nevertheless made from algae genes altered using synthetic biology methods. These purportedly good intentions could very well open the door for legitimizing a risky and little-tried technology which may then be applied to possibly nefarious aims. Following massive protests by NGOs Ecover shelved these plans; while an "open public dialogue" has been initiated.

It has become evident that risks and problems inherent in research and more so in the industrial application of synthetic biology are enormous and multifaceted. Medical applications obviously require special consideration and shall not be dealt with in this article—even if they constitute a central sector of current research. With regard to research and application of synthetic biology in manufacturing of fuels, plastics, flavorings and other products in the bioeconomy context, the following problem areas are decisive:

Synthetic biology creates self-replicating organisms destined to be released into the environment or used in supposedly closed-off labs resp. factories. Security risks in both cases are enormous. Possible consequences of contamination with synthetically produced organisms or genetic material for humans and the environment are literally impossible to assess.

Products produced using synthetic biology methods (like additives in food or cosmetics) require no labeling at this time; they are rated "natural". This constitutes a case of massive consumer deception. Entailed is also widespread loss of employment in southern areas where millions of people will be unemployed once synthetic vanilla flavoring or a synthetic substitute for coconut oil will supplant present-day agricultural produce. In the Philippines for instance 25 million people are directly or indirectly dependent on coconut production.

But even synthetic biology cannot create something out of nothing. Bacteria, algae and microbes need to be fed. Up to now this was mainly met by sugars. Therefore many synthetic biology companies run large sugar cane outfits or sugar refineries—most in Brazil. Sugar, however, like other agricultural produce—is traded on world markets, subject to rules of supply and demand. High demand for sugar leads to price increases and eventually to the conversion of many previously agriculturally-used areas into sugar plantations, or the clearcutting of wooded areas and causes high stress on biodiversity. It remains to be seen if the final balance for palm seed oil substitutes will be as positive as predicted. And even if the input for synthetic biology can be changed from sugar to wood or any other type of so called biomass in the future, the fact remains that biomass production requires land, water and other natural resources, thus turning into direct competition for food production for a steadily growing global population.

Doubtless companies investing large sums of money into synthetic biology research aimed at benefitting the fossil fuel industries are well aware of this. This trend is especially remarkable given that many synthetic biology companies have explicitly presented themselves as alternatives to a continued fossil future. This involves fracking as a means for methane production from natural gas and crude oil extraction as a sugar substitute or other types of biomass. This would mean an enormous increase in the value of this gas, something these companies would surely appreciate given the low price of crude oil. Additionally synthetically produced microorganisms could be utilized in the exploitation of hard to reach oil and gas reserves. While easily reached oil reserves are being depleted, there is more and more "residual oil" left for companies to exploit deploying various methods. These methods are called "Enhanced Oil Recovery". One of these methods gaining more and more ground is MEHR ("Microbial Enhanced Hydrocarbon Recovery"). Microorganisms are "programmed" and injected into a wellbore where they form certain chemicals in order to extract oil resp. get it ready for extraction (and later transport). There have been more than 300 known test sites. Companies like BP, Shell and Statoil are investing in these methods.

Besides the fundamental cultural shift towards a discussion of biomass and bioenergy instead of talking about bacteria, plants, trees or soils as living entities with their specific integrity at stake, one problem is that these synthetic biology companies secure patents for themselves which in the final analysis would be "patents on life".

3 Patents on Life: A Challenge to Integrity

Patent applications like the ones by the US Recombinetics company which specializes in the application of novel genetic engineering processes in animals, show where the development is heading. The parent company of Recombinetics is the British Genus Company, the worldwide largest animal breeding company. Recombinetics applies for patents on animals for agricultural purposes as well as pharmaceutical research. Their patents list beef cattle with more muscle mass and pigs modeling diseases. In doing so, they apply for patents rendering animals genetically sterile. This serves the purpose of keeping farmers from breeding these patented animals further. In 2014 and 2015 several of these patent applications were submitted for both primates and even humans (see table below).

At the end of 2015 the announcement was made that the company wants to market its patented, hornless beef cattle. It can be assumed that soon these patented animals are expected to yield monetary returns.

While many protagonists of synthetic biology advertise on talk shows, in colorful folders and on their websites that their DNA-computer-designed "BioBricks" like some sort of Lego brick can be cut and recombined endlessly, real biological science is heading into another direction. Genetics has taught us during the past years that there is a lot we still do not know. DNA has turned out to be much more complex than previously assumed. It has its own integrity. Information involves various and interlocked genes. There is no way of predicting what will happen with all this snipping and recombining as if it simply involved a large building set.

But why are so many companies and governments predominantly in the US, Great Britain, France, the Netherlands, Denmark, Switzerland, Germany, Canada, China, Brazil, Japan and Australia investing so much money? Even large private foundations like the Bill & Melinda Gates Foundation, the Sloan Foundation and the Gordon and Betty Moore Foundation invest prodigious sums. Could this constitute a financial bubble? This remains an object of speculation. Much of what the great bioeconomy vision based upon human-controlled synthetic biology promises will never be realized. So far, many products having been introduced to the market have proven their disadvantageous impact upon food safety, living standards, biodiversity and climate change. They have proven that there is a definite need for keeping a sharp eye on research and testing in the context and name of bioeconomy or, questioning it more profoundly, there is a need for a broad public discussion about benefits and risks entailed in this whole approach. (see for this chapter: Fatheuer et al. 2015).

Table Selection of patent applications by US-Recombinetics company

Number of international patent application and content	Claims
WO 2011100505 Pigs as Guinee Pigs. The animals are bred to have genetic defects responsible for myasthenia, baldness or arteriosclerosis. According to the patent application the animals may show signs of "damage due to hypercholesterolemia and arteriosclerosis".	Pigs with abnormal genetic structures connected to lipoprotein receptors, myasthenia and baldness.
WO 2012116274 Beef cattle and pigs showing increased muscle mass.	Process using the TALEN nuclease to alter genetic disposition for muscle growth. Cells, embryos etc. of the following types of animals: "Ungulates, pigs, cattle, fishes, rabbits, livestock" (Claim 28) as well as the genetically altered animals thus produced.
WO 2013192316 Among others hornless cattle, cattle breeds like Wagyu (Japan) and Nelore (India) showing more muscle mass and increased milk yield. Animals with genetic defects suffering from cancer and circulatory diseases as animal models for pharmaceutical research.	Processes using the TALEN nuclease to genetically alter cells, clone embryos and transfer them to surrogate mothers. Cells from among others "pigs, cattle, sheep, goats, chickens, rabbits, fish, zebra fish, dogs, cats, mice, rats and laboratory animals". (Claim 28) as well as any genetically altered animals thus produced.
WO 201470887 Farm animals that do not become sexually mature can be fattened longer. Farmers are unable to use these animals for breeding. If breeding is desired, hormone injections are required.	Use of TALEN nuclease, zinc finger and CRISPR blocks gene necessary for sperm cell production. Genetically altered animals like "non-human vertebrates, non-human primates, cattle, horses, pigs, sheep, chickens, birds, rabbits, goats, cats, laboratory animals and fishes". (Claim 21)
WO 2015030881 Transfer of genetic variations in livestock and laboratory animals with the help of nucleases.	Application of nucleases like zinc finger, TALEN, and CRISPR in cells from "non-human vertebrates, primates, pigs, horses, sheep, goats, chickens, rabbits, fishes, dogs, mice, cats, rats or laboratory animals" as well as the genetically altered animals produced this way.
WO 2015168125 Animals which have been altered genetically multiple times—the text names up to 25 genes—to be genetically altered. Among others, the process is designed for application in livestock and animals bred for organ donorship.	Methods using nucleases and zinc finger, TALEN and CRISPR for multiple genetic cell alterations; including human cells. Animals genetically altered multiple times including higher primates (humans are part of this) and livestock.

Then (2016)

4 Integrity at Risk: The Need for a Bioethics of Integrity

Nowadays mankind is preoccupied with long-term and far-reaching risks inherent in technological innovations such as biotechnologies, nanotechnologies and fracking. Added to these are risks that deeply and globally change societies given that they usually entail profound political consequences like terrorism, climate change, nuclear phase-out, renewable energy sources, digitalization. These systemic risks which correlate with interconnected problem areas whose consequences with regard to extent, depth and time-frame are the basic issues of post-modern societies. (cf. Renn et al. 2007, pp. 161–191, 166.).

Risks inherent in new technologies obviously threaten integrity. Whenever integrity is harmed, there is a need for ethical justification. However, so far there has been no adequate ethical or cultural discourse on this topic. Thus the need for a bioethics of integrity is apparent and urgent, including topics of political governance with integrity.

Especially synthetic biology obviously has the potential of changing the profile of environmental biodiversity, given that it might wreak changes in general evolutionary patterns. Thus a bioethics of integrity is called for especially with reference to the latest developments in biotechnologies.

Integrity on the one hand means "Intactness" of a life form. Many procedures in animal husbandry have been harmful, violating the intactness of animals (tail docking, tooth clipping), not to mention starting with gene-editing or feeding animals with biomass created using synthetic biology. On the other hand, in ethics, integrity is regarded as the honesty, truthfulness or accuracy of one's actions. A person or organization has ethical integrity if individual actions, beliefs, and principles all derive from a core group of values deeply rooted in a specific culture or set of (religious) beliefs. Synthetic biology and gene-editing will bring forth a completely new culture of dealing with life and living beings which in no way constructively resonates with the inherited cultural mindsets prevalent at the beginning of the twenty-first century. Who will address the cultural revolution which is a prerequisite of the coming technological revolution putting integrity at risk?

Industrial agriculture and the food industries are two sectors where modern biotechnology will open up new ways of doing business. They are two economically, ecologically and socially central fields of future well-being or an integral evolution. But who is responsible for what? Farmers, the chemical industry, biologists, governments, commerce—who is responsible for jeopardizing the integrity of (all) life forms?

5 Risk Governance and Framework

A modern answer to this question lies in the field of risk assessment and risk management.

The International Risk Governance Council (IRGC) is an independent non-profit foundation which aims to help improve the understanding and management of risks and opportunities by providing insight into systemic risks that have impacts on human health and safety, on the environment, on the economy and on society at large.

As a science-based think tank and neutral collaborative platform with multidisciplinary expertise, IRGC's mission includes developing concepts of risk governance, anticipating major risk issues, and providing risk governance policy advice for key decision-makers.

Drawing upon international scientific knowledge from both the public and private sectors, IRGC champions ignored, neglected and emerging issues and can help building bridges between science and policy in today's challenging governance environment.

The IRGC (2005) four-phase concept of risk governance has been designed to allow for better climate risk predictions and their handling. They propose a comprehensive risk-handling chain which could be adopted for biological or environmental risks inherent in synthetic biology as well. Risks in this concept are not determined technically or science-oriented, they emerge from interplay between human behaviors and natural reactions. A four-tier process has been designed comprising all significant aspects of an effective way of dealing with risks, while never losing sight of public concerns. The publication also establishes a consistent terminological framework to aid research into risk reduction and reduce confusion. The concept also serves as an instrument for evaluating good governance, thus assuring a complete, effective, efficient and socially acceptable way of dealing with risks (IRGC 2005, p. 17).

Preliminary phase: What does an "integrative framework" mean? Under ideal conditions the first phase of the steering process would be the pre-assessment. Its main focus is the framing of the problem, i.e. conceptualization and limitation of risk under consideration. Participants have to determine which prerequisites and selections will apply and how the risks under consideration can be compared (IRGC 2005, p. 24f). Frequently frames are tied to cultural or socio-geographical contexts. Framing determines if a phenomenon need be considered a risk and if so, which causal functional chains need to receive closer inspection, and which facts need integration or exclusion. In this phase questioning of stakeholders is helpful in order to become acquainted with their views and in dialogue with these stakeholders determine individual risk research and management concepts. Specific cultural and geographic contextual determinants have to be taken into consideration.

In addition to framing the preliminary phase has additional procedural steps (IRGC 2005, p. 24 ff., Ad-hoc-Kommission 2003; Renn and Walker 2008, p. 48) like

- Institutional procedures for early risk warning and information on undesirable developments relayed to risk management institutions—like an early warning system for possible development of damage.

- Generally valid guidelines allowing for the early establishment of consistent and replicable risk response procedures—for instance agreement on central indicators and an assessment process.
- Screening allowing for early determination of risks and methods and scientific steps required—for instance a quick summary process for the early detection of possible supply bottlenecks during changeovers to renewable energy sources.
- Scientific procedures and techniques (scientific conventions) used in the determination of risks—for instance an agreement on the validity and significance of assessment processes applied in climate prognoses.

The second phase of the IRGC model deals with scientific declaration of risks. It deals with risk assessment and identification of popular concerns. (Risk perception and concern assessment). (IRGC 2005, p. 26ff). Generally speaking, physical risks and popular concerns have to be analyzed using best scientific methods and—if possible—quantified. The results of this scientific diagnosis will become part of the comprehensive risk assessment. Compilations of risks like health and environmental risks, economic wellbeing and societal stability have to be supplemented with analyses of risk perceptions and opinions of important social groups as well as the general population impacted by these risks. It is important to make efficient use of the pool of knowledge and experiences available. It is also important to include the dimension of time. Conflicts often arise due to one side viewing risks short-term, while the other takes the long-term view. Additionally there is the question of local limits of negative impacts: Are we dealing with an area of a county, a country, a continent as a whole or the entire world?

Risk assessment: Determination of acceptability. Phase three starts as soon as all data has been assembled. Data is compiled, interpreted and evaluated. According to the IRGC model this is a two-step process: Characterization of risk and risk assessment (IRGC 2005, p. 36ff). The main focus is categorizing the risk as acceptable, requiring correction or intolerable. To judge acceptability requires introducing possible harm or benefit of the activity in question, such as energetic service performance as well as global and health risks caused by power plants burning fossil fuels.

Including uncertainties in the evaluation is especially difficult. If it is known what consequences a certain risk will have, probabilities can be included and quantified. This is impossible when dealing with unknown or hard to evaluate risks. In these cases, assessments can only be based on subjective ideas versus tenuous consequences. Mathematical quantification is barely possible when dealing with particularly uncertain consequences. In these cases a risk dialogue with stakeholders and pertinent scenarios will have to serve. Factors difficult to quantify like environmental protection, biodiversity and stability of ecological systems have to be included here.

The more doubtful the risk and the more controversy surrounding it the more difficult will it be to weigh risk and costs and make a comparison. For example: climate protection is a case in point. When assessing the consequences of a business-as-usual-scenario in comparison to an effective climate protection scenario, the expenses incurred for doing nothing have to be juxtaposed with those incurred for effective climate protection. And it is not only money that is at stake, but ecological, social and cultural harm. These "soft" consequential damages cannot be measured objectively; they require instruments of empirical social science like surveys of stakeholders of eco-systemic services, systemic analyses of documentations and eco-metric evaluation procedures. The IRGC recommends a risk dialogue for administrative representatives, members of the sciences and stakeholders affected in order to compare hard and soft outcomes of the analysis. This process will end in a judgment determining which climate-consequence-scenarios will be acceptable resp. tolerable. (Fairman 2007; Renn and Walker 2008, p. 149ff).

Risk management and risk communication. The fourth phase deals with risk management. Here concrete measures or strategies will be selected in order to forestall a non-tolerable risk or lower it to such an extent as to make it acceptable (IRGC 2005, p. 40ff). The IRGC implements analytical methods of decision-making. When dealing with risk management, measures of risk reduction have to be implemented, which result in suitable concepts for different levels of complexity, insecurity and ambivalence of initial risk.

All four phases require intensive risk communication plus a discursive risk dialogue. Contrary to earlier directions for risk treatment—such as the National Research Council's (NRC) of 1983—IRGC risk communication provides for a continually advancing process, starting with the preliminary phase all the way to risk management. Demanding quick and comprehensive communication is not only due to reasons of democratic decision-making, it also enriches management processes. (cf.: IRGC 2016).

6 Who Draws the Line? – Risk Ethics and a Governance of Integrity Are Needed

In revolutionary times—and the twenty-first century, also called the "Age of Biology", certainly is such a time—it is difficult to rely on sound and just legal systems. Indeed despite the "idealized" IRGC model outlined above, at the moment there is a practical lack in political governance of the biotechnologies or of bioeconomy as the new paradigm for what is called a sustainable industrial revolution.

There are some NGO bodies watching these new technological powers and new business opportunities connected to them, but there is no political discourse on any kind of formal regulation of gene-editing or synthetic biology with respect to the integrity of people, plants and animals. To ensure the "future of biology" and introduce products and services to their economies and in spite of assisting in the creation of frameworks for ethical discourse, politicians make available large sums of money for research purposes. There are research policies, but there is no ethical line with respect to what changes may or may not be wrought by genome-editing and other techniques with regard to integrity.

Therefore, a modern risk ethics is needed. Risk ethics allows making informed judgments of morally relevant present-day and future risks. An ethical risk discussion based upon applicable ethical principles is required when defining the compatibility of social and ecological measures. The high complexity and interconnectness of present-day technical interventions and large-scale projects require these to be "error-friendly". This means that technical innovations or interventions have to be stopped in case of negative ramifications and that no irreversible harm will result. Especially the second aspect, reversibility, is not given with technologies fraught with very high and/or unknown risks like nuclear energy or many biotechnologies. This makes (independent) scientific risk research extremely important. Only risk research, based on the IRGC model but grounded in real processes of broad public ethical dialogue, is able to define and open an ethical discourse on verifiable criteria allowing for the demonstration of manifold positive or negative outcomes of technical solutions as well as consequences for the environment and societies hitherto unknown, which are spatially and chronologically difficult to plot. It has been the rule in conflict situations that risk ethics must be able to employ a societal discourse in order to appropriately weigh benefit and cost and properly communicate the outcome. For this end ethical basic principles like responsibility, prevention and (generational) justice have to be considered. A measure is only ethically sound if the negative outcomes and side effects have been reduced to the smallest possible amount. If the impact remains very high once risks and negative effects have been reduced, proof will be required why foregoing the measures or the technical applications would have gravely negative outcomes.

This does not hold true at the moment. Especially in the biotechnology area much has been researched and implemented before the cost of relinquishing the plan—for the general public, mind you, not for business—has been properly discussed. Therefore it is not enough to supply risk research and technology assessment with ethical instruments in order to evaluate decisions. What is needed is a general public discussion concerning the risks people are willing to run before binding political instruments like laws, regulations etc. will be enacted.

A modern risk ethics calls for a strong governance structure with regard to new technologies. A good example for what is needed is the Biosafety Protocol (CBP):

The Cartagena Protocol on Biosafety to the Convention on Biological Diversity is an international agreement on biosafety as a supplement to the Convention on Biological Diversity effective since 2003. The Biosafety Protocol seeks to protect biological diversity from the potential risks posed by genetically modified organisms resulting from modern biotechnology. I would like to end with a reference to an extensive and leading-edge text on biosafety by Wikipedia:

Biosafety is the prevention of large-scale loss of biological integrity, focusing both on ecology and human health. These prevention mechanisms include conduction of regular reviews of the biosafety in laboratory settings, as well as strict guidelines to follow. Biosafety is used to protect from harmful incidents. High security facilities are necessary when working with synthetic biology as there are possibilities of bioterrorism acts or release of harmful chemicals and/or organisms into the environment. A complete understanding of experimental risks associated with synthetic biology is helping to enforce the knowledge and effectiveness of biosafety.

Biosafety is related to several fields:

- In ecology (referring to imported life forms from beyond ecoregion borders),
- In agriculture (reducing the risk of alien viral or transgenic genes, genetic engineering or prions such as BSE/"MadCow", reducing the risk of food bacterial contamination)
- In medicine (referring to organs or tissues from biological origin, or genetic therapy products, virus; levels of lab containment protocols measured as 1, 2, 3, 4 in rising order of danger)
- In chemistry (i.e., nitrates in water, PCB levels affecting fertility)
- In exobiology (i.e., NASA's policy for containing alien microbes that may exist on space samples. See planetary protection and interplanetary contamination), and
- In synthetic biology (referring to the risks associated with this type of lab practice).

The international Cartagena Protocol on Biosafety deals primarily with the agricultural definition but many advocacy groups seek to expand it to include post-genetic threats: new molecules, artificial life forms, and even robots which may compete directly in the natural food chain.

Biosafety in agriculture, chemistry, medicine, exobiology and beyond will likely require application of the precautionary principle, and a new definition focused on the biological nature of the threatened organism rather than the nature of the threat.

When biological warfare or new, currently hypothetical, threats (i.e., robots, new artificial bacteria) are considered, biosafety precautions are generally not sufficient. The new field of biosecurity addresses these complex threats.

Biosafety level refers to the stringency of biocontainment precautions deemed necessary by the Centers for Disease Control and Prevention (CDC) for laboratory work with infectious materials.

Typically, institutions that experiment with or create potentially harmful biological material will have a committee or board of supervisors that is in charge of the institution's biosafety. They create and monitor the biosafety standards that must be met by labs in order to prevent the accidental release of potentially destructive biological material. (Wikipedia, Biosafety). Little need be added to this practicable, ethical protocol on integrity and the preservations of life forms. It might be worthwhile to keep in mind that the Cartagena Biosafety Protocol unequivocally states that products from new technologies must also be based on the responsibility principle and allow developing nations to balance public health against economic benefits. It will for example let countries ban imports of genetically modified organisms if they feel there is not enough scientific evidence that the product is safe and requires exporters to label shipments containing genetically altered commodities such as corn or cotton.

One thing is clear: Despite all the activities listed in the CBD, a new approach and governance are needed which are tailor-made to address the new dangers posed by synthetic biology to the integrity of all life forms before it is too late. But who will take the lead?

References

Ad-hoc-Kommission (2003) Neuordnung der Verfahren und Strukturen zur Risikobewertung und Standardsetzung im gesundheitlichen Umweltschutz der Bundesrepublik Deutschland. Abschlussbericht der Risikokommission. Bundesamt für Strahlenschutz, Salzgitter

Fatheuer T, Fuhr L, Unmüßig B (2015) Kritik der Grünen Ökonomie. Oekom, München

Fairman R (2007) What makes tolerability of risk work? Exploring the limitations of its applicability to other risk fields. In: Bouder F, Lofstedt RE, Slavin D (eds) The tolerability of risk: a new framework for risk management (Earthscan, London), FSA, Food Standards Agency, London, pp 119–138

IRGC (2005) Risk governance – towards an integrative approach. Retrieved from https://irgc.org/wp-content/uploads/2012/04/IRGC_WP_No_1_Risk_Governance__reprinted_version_3.pdf

IRGC (2016) https://www.irgc.org/wp-content/uploads/2016/11/Basabe-P.-2016.-Comparing-and-contrasting-approaches-to-risk-governance.pdf

Renn O, Dreyer M, Klinke A, Schweizer, P-J (2007) Systemische Risiken: Charakterisierung, Management und Integration in eine aktive Nachhaltigkeitspolitik. In: Jahrbuch Ökologische Ökonomik, 2007, Bd. 5, Soziale Nachhaltigkeit, pp 161–191, 166

Renn O, Walker K (eds) (2008) Global risk governance: concept and practice using the IRGC framework. Springer Science & Business Media, Heidelberg

Then C (2016) Gentechnik, Patente und die Tierversuchsindustrie. Eine Studie im Auftrag der Fraktion Bündnis 90 / Die Grünen im Bundestag. https://www.gruene-bundestag.de/fileadmin/media/gruenebundestag_de/themen_az/agrar/PDF/160809-analyse-gentechnik-tiere-ethik.pdf

Wikipedia, Biosafety. https://en.wikipedia.org/wiki/Biosafety

WZW (2016) Lehrstuhl für Biotechnologie der Nutztiere, http://btn.wzw.tum.de/index.php?id=44)

From Ecosystem Services to Ecological Solidarity

Agnes Michelot and Anna Aseeva

1 Introduction

In 2016, anthropologist Veronica Strang suggested that 'justice'—more precisely, 'relational justice'—is when we recognise, appreciate and uphold *value* in *others* (Strang 2016).[1] All-embracing discussions of 'Other' and 'Others' have been extensively narrated elsewhere, so, there is no need for us to paint their full canvas here.[2] Moreover, what is of great interest for us here is not that much the 'Other' or 'Others', but rather questions of justice against the background of *recognising, appreciating and upholding 'value', in its broad sense, of others*. In this chapter, we will attempt to enquire how and why our society arrived at the point of monetary valuation, and the following contractualisation and commodification of natural functioning of ecosystems.

We undertake a study of how much an economic valuation of ecosystems is encouraged and imposed in and through a broader mainstream neo-classical economics vision of nature and biodiversity. This study aims at answering the following central question: how the interests of nature could be (re-)thought and represented beyond the economic valuation of ecosystems, if at all?

We set the stage with an overview of key conceptual elements of the above interface (2). We continue with the analysis of the discourse, politics and principle of ecosystem services (3) and potential effects of that concept in theory and practice of law (4); as well as of possible ways of contestation and resistance that include the

[1]Strang (2016), pp. 263–278. Cited in Kopnina (2016), p. 181.

[2]To quote just a few, see: in philosophy, Hegel (1977). Hoffmeister (1977); E. Husserl, *Cartesian Meditations: An Introduction to Phenomenology*, 1931; in ethics, Lévinas, E.. *Otherwise than Being or Beyond Essence*, 1974; in postcolonial studies, McClintock (1995).

A. Michelot (✉) • A. Aseeva
University of La Rochelle, La Rochelle, France
e-mail: agnes.michelot@univ-lr.fr; anna.aseeva@graduateinstitute.ch

L. Westra et al. (eds.), *The Role of Integrity in the Governance of the Commons*,
DOI 10.1007/978-3-319-54392-5_3

ecological solidarity—a principle that emerged in French law theory and practice (5). Drawing on lines of ecological solidarity, we close the chapter with a discussion on a further possibility of just relationship between humans *and* other species *and* ecosystems as a viable alternative to monetary valuation and continuous exploitation *of* ecosystems *by* humans (6). Lastly, we offer our concluding remarks.

2 From Ecosystems to Ecosystem Services: Towards Monetarisation of Ecology?

'Ecosystem' is generally defined as a dynamic grouping of a living ('biotic') community and its non-living ('abiotic') environment.[3] The ecosystem diversity thus means diversity of such groupings.

The ecosystems approach is a concept, which is currently widely used in different fora.[4] It consists of protection and preservation of rare or fragile ecosystems as well as the habitat of depleted, threatened or endangered species and other forms of life, but emphasises the protection and conservation of an ecosystem *as a whole*, rather than certain sectors or species.[5]

In scientific literature, the theory of ecosystem services was introduced and developed in the late 1980s by the ecological economics movement as part of a new approach to study the environment (de Groot 1987; Costanza and Daly 1992; Jansson et al. 1994; Costanza et al. 1997; Daly 1997; Gómez-Bagghetun et al. 2010; Froger et al. 2012). The crux of these developments was twofold: firstly, to highpoint the economic benefits that humans can extract from nature and particularly from the functioning of its ecosystems; and secondly, and consequently, to propose concrete tools that would quantify those economic benefits as well as 'losses', and thus the monetary value of ecosystems.[6]

Today, functional definitions of ecosystem services comprise "direct or indirect contribution to human well-being" (TEEB 2010, p. 33) or else "the benefits people obtain from ecosystems" (MEA 2005, p. 53; Carpenter et al. 2006).

The pluralistic nature of the objectives of the economic valuation of ecosystem services, which became a cornerstone of the mainstream scientific and political system for the understanding of and working with sustainability issues (Prevost et al. 2016), yet triggers debates in both academic literature and public sphere.

---

[3]Maris (2016), p.75.

[4]EU Marine Strategy Framework Directive; Convention for the Protection of the Natural Resources and Environment of the South Pacific Region (adopted 24 November 1986, entered into force 22 August 1990) 26 ILM 38 (Nouméa Convention); Adoption of the Paris Agreement. Proposal by the President.(UNFCC)(12December2015) FCCC/CP/2015/L.9/Rev.1 (Paris Agreement).

[5]For marine ecosystems, see eg. CBD Aichi Target 6; UNGA Resolution 44/225 (22 December 1989) A/RES/44/225; Churchill (2015), p. 11; Barritt and Viñuales (2016), p. 47.

[6]See eg. Maris (2016), pp. 15–16.

Various debates and criticisms reveal both difficulties and ambiguities of the valuation exercise.

3 Ecosystem Services: A Questionable Policy Instrument

Theoretical ideas on economic valuation of ecosystem services offered plural opportunities for conceptual innovations, thus 'fresh' and original grammars of environmental economics with an underlying rationale of economic efficiency. Since the 2000s, the narrative of ecosystem services therefore fully integrated a popular discourse in the mainstream political arena elsewhere (Fitoussi and Laurent 2008; Chevassus-au-Louis et al. 2009).

One of the most striking examples could be provided by the CBD Nagoya Protocol. Namely, at its 10th Conference of the Parties, it set 20 Aichi Targets. In their relevant parts, the targets essentially claim that biodiversity should be valued in accounting terms in order to be "integrated into national and local development and poverty reduction strategies and planning processes and are being incorporated into national accounting..."[7] The most relevant targets mention a utilitarian valuation of biodiversity through *services* that it can provide to humans (targets of the group D).

A political readiness to value and derive, even if sustainably, benefits that biodiversity 'offers' to mankind hence seems to reach its heights in the principle and politics of 'ecosystem services'. Today, such principle and politics are integrated in various regulatory and legal *conservation* strategies and instruments at all levels (often through the ecosystem approach, sometimes without expressly mentioning the 'services' angle).[8] They thus systematically directly assimilate the natural role of biodiversity[9] to its pecuniary values, and oftentimes even conditioning and justifying its conservation by such values.[10]

Here, a simple yet important normative question could have already been arisen: how a *monetary* valuation can assess *non-market* values—especially those that

[7]https://www.cbd.int/sp/targets/#GoalA.

[8]Convention for the Protection of the Natural Resources and Environment of the South Pacific Region (Nouméa Convention) [1990] 26 ILM 38; UN-sponsored Millennium Ecosystem Assessment (2005), with a reference to classification for ecosystem services; Adoption of the Paris Agreement, Proposal by the President(UNFCC) [2015]FCCC/CP/2015/L.9/Rev.1 (Paris Agreement). For marine ecosystems, see eg. DIRECTIVE 2008/56/EC OF THE EUROPEAN PARLIAMENT AND OF THE COUNCIL establishing a framework for community action in the field of marine environmental policy (Marine Strategy Framework Directive) [2008] OJ L164/19; CBD Aichi Target 6; UNGA Resolution 44/225 (22 December 1989) A/RES/44/225.

[9]That is, developing, streaming and sustaining adaptability and self-protection properties of any forms of life.

[10]*The Intergovernmental Science-Policy Platform on Biodiversity and Ecosystem Services (IPBES)* provides the best example of the above through its very title.

cannot be quantitatively assessed, so, measured—of a subject? This question was however never answered and not even really asked in the relevant economic literature or the public debate.

The notion of service, however, indeed presents a number of difficulties—in both economics and law. Virginie Maris, for instance, pointed out that an ecosystem, entity, function or set of functions simply could not always correspond to, or present a service.[11] She thus underlines that certain functions and environments simply could not represent any kind of commodity for humans. On the other hand, the ecosystem service approach may present some normative and relational difficulties from the distributive justice angle (Sikor 2013; Roche et al. 2016). Namely, *benefit* implies an existence of *beneficiaries*, but also, drawing on basics of distributional justice, of a potential *distributor* of those benefits, who would also ideally designate the exact beneficiaries.[12]

An issue relating to economic assessment, yet going beyond a blunt cost-benefit analysis is that of value—that is, not necessarily and not only an economic notion of value, but rather its broader philosophical and ethical implications. The latter somewhat relate to the nature of the value. In this regard, some developments taking place in international environmental law over the last decades suggest significant ruptures, also triggering debates in the political arenas.

4 Concept of Ecosystem Services in Theory and Practice of Law

In order to grasp the impact of the widespread introduction of ecosystem services in the field of environmental law, it is important to remember the ethical values and legal principles this branch of law draws on. It is worth reminding that this field of law is intended to take nature into consideration in human activities. Historically, ecology has long been considered as being outside law, in the sense that ecosystems and natural habitats had no special legal status. It is only in the beginning of the twentieth century that laws applicable to nature or to its components started to emerge. International environmental cooperation started in the early twentieth century, when the First International Congress for the Protection of Nature was held in Paris.[13] For the last fifty years or so, the purpose of some instruments within the core of international environmental law became the one of identifying, regulating and managing processes that affect biological diversity.

[11]Maris (2014), p. 31.

[12]Distributive justice posits that a certain good or benefit, is being distributed by a distributor equally between all entitled recipients; and equals be treated equally. Wissenburg (1993), p. 6.

[13]1913: First International Congress for the Protection of Nature, Paris, and creation of the Consultative Commission for International Protection of Nature, inspired by Paul Sarasin (famous Swiss environmental activist)—World War I impeded any further developments.

The 1972 United Nations Conference on the Human Environment that took place in Stockholm has led States to advance ecological ambitions beyond their national concerns. In this context, a comprehensive approach has been developed, attributing a value to nature that is independent of its direct relevance to humans. From this point of view, the international conventions adopted after the Stockholm Conference represent particular interest for our topic. For example, the 1979 Bonn Convention on Migratory Species acknowledges in its preamble that "wild animals (...) are an irreplaceable part of the earth's natural systems." The Bern Convention on the Conservation of European Wildlife and Natural Habitats 1979, stated: "wild flora and fauna constitute a natural heritage of aesthetic, scientific, cultural, recreational, economic and intrinsic value..."

The World Charter for Nature adopted 28 October 1982 by the UN General Assembly is based on the intrinsic value of nature, while referring to the moral dimension of the recognition of such value to all living beings: "Every form of life is unique, warranting respect regardless of its worth to man, and, to accord other organisms such recognition, man must be guided by a moral code of action."

The representation of the intrinsic value of nature was, however, seriously challenged since the 1992 Rio Summit on Sustainable Development. Indeed, these negotiations have led to the adoption of the United Nations Framework Convention on Climate Change (UNFCCC) and the Convention on Biological Diversity (CBD). Notably, in the Preamble of the CBD, State parties declared an awareness of "the intrinsic value of biological diversity and of the ecological, genetic (. . .) values of biological diversity and its components."

Overall, in its actual Preamble, the CBD promotes (1) the sustainable use of biodiversity components, (2) the conservation and (3) the fair sharing of benefits of the genetic resources in areas under national jurisdiction. Hence, CBD's third objective (benefit-sharing, so economic values) relate and complement its first and second objectives—conservation and sustainable use (non-economic values) of biodiversity.[14] On the other hand, it is generally viewed that since the beginning of negotiations, the CBD was about resources, so, state sovereignty, rights, and the like, or, even following some more conservative views, that it is a "realistic" instrument respectful of state sovereignty regarding their natural resources.[15]

The idea that it was necessary to consider the erosion of biodiversity in terms of *net* economic losses gradually took precedence over the recognition of *intrinsic* value, and not only within the CBD, but in various arena.[16] However, it is argued that reasoning in terms of 'net' losses, especially as opposed to losses *tout court,* means also admitting, at least implicitly, that any damage to nature and biodiversity could be compensated and the final 'net' account could thus look acceptable.[17] Moreover, and consequently, unlike, for example, 'green accounting', this type of

[14]Morgera et al. (2014), p. 3.

[15]F.Francioni, Foreword to Morgera et al., ibid., p.xiii.

[16]See eg. European Commission, EU Horizon 2020.

[17]See eg. Maris (2016), p.17.

compensation logic seems to take for an analytical unit not the economic utility of biodiversity, but directly its *ecological* value.

At that point, a significant shift in global *legal* discourse occurred, even though research on the assessment of ecosystem services was just about to develop (Doussan 2009). Indeed, environmental law has traditionally played a role of governing *humans* and their relationships, including the moral dimension, whatever the representation of the human-nature relationship was. This field of law was built on the prevention of the damage caused to the environment and therefore on an essentially regulatory approach that consists of the provision of a framework for activities and of modalities of precaution and prevention, as well as reparation in the event of environmental damage. Therefore, the social construction of human-nature relationship[18] potentially entails a major legal rupture and has two important consequences. Namely, two critical outcomes of such a system are: (1) the benefits derived from nature are of value to humans but, above all, are a condition of our survival and/or well-being; and (2) the objective of *human* well-being and interest becomes the major motivation for the protection of ecosystems.

The recognition of the concept of ecosystem services as we know it today thus marks *the abandonment of the intrinsic value*, and potentially the challenge to all mechanisms of environmental law based on the obligation of *protection* organised around the basic principles of *prevention* and *precaution* (Naim-Gesbert 2014; Kourislky and Geneviève 1999; Prieur et al. 2016), as opposed to the logic of *substituting* everything in nature—that is, damaging it and then 'compensating' (Callon 1998).

Moreover, current concept of ecosystem services is based on a normative rationale where the relationship between the legal subject and object are overturned. Indeed, the type of legal relationships relating to the environment is no longer based on identification, protection or even the support of ecological functions that, by definition, belong to no one and are not subject to economic evaluation. Conversely, in the current construct, ecological functions of ecosystems now have to support the well-being of humans. The notion of a service implies, in particular, the identification of a supplier and a beneficiary of the service. This opens the door to all forms of contractualisation and further commodification, but for which the environmental protection is, in legal reality, no longer the object, although contractual techniques may also be used to achieve environmental goals (Hautereau-Boutonnet 2015). In other words, the evolution of the relationship between mankind and nature, which leads to the latter being assessed and measured according to the benefits derived from ecosystems, seems to suggest moving away from the regulatory approach and an opening of environmental law to a 'pragmatic' discourse and practice of economic assessment of services provided by ecosystems.

[18]Note that this construct was implicitly defended in the Millennium Ecosystem Assessment (2005), and reaffirmed in the 2010 Nagoya Protocol on Access to Genetic Resources and the Fair and Equitable Sharing of Benefits Arising from their Utilization to the Convention on Biological Diversity.

Law is then potentially reduced to a rationale of management and appropriation of ecosystem services that simply allows the creation of new relevant legal tools and mechanisms to better serve this economic model become mainstream (Hervé-Fournereau and Langlais 2013).

It must still be remembered that law is an instrument of social control geared towards justice (Apostolidis et al. 1993). Consequently, its orientation in terms of principles, mechanisms and instruments that it mobilises is loaded with particular symbolic significance.

5 The Way Forward

Looking at the paradigm of ecosystem services through the prism of biodiversity and broader relational justice, we argue that the problem is that today, the relevant vision of *the just* within the discussed distributive models relating to ecosystem services may induce a *more just* commodification of biodiversity, but clearly lacks the critical workings to contest commodification *per se*.

The only model that today seems to attempt to strike a balance between the two paradigms, i.e., the intrinsic value of biodiversity and economic valuation of biodiversity, is the CBD's Nagoya Protocol. But Nagoya Protocol has a number of critical flaws, especially regarding workings of economic valuation and further contractualisation of nature and biodiversity. They seem to be crystallised in the Nagoya Aichi Targets, streaming into national legislations, standards, and policies the dynamics of utilitarian and accounting use of ecosystems. Otherwise, these important shortcomings may also be rooted in the fact that, as a part of the CBD, the Nagoya Protocol is limited by, and indeed bound by state sovereignty and frontiers.

We may therefore have a general political impulse to be today interested in the idea of *commons* because, first of all, the number of strong arguments against both *property* and *sovereignty* is growing. What may come with the aforesaid arguments is the idea of the relation between the notions of *commons* and *community*. However, the first concrete problem—both theoretical and analytical—that immediately follows such an assumption is the one of representation. That is, today, most of relevant discourses and practices are still strongly linked to the state-form universality. CBD is only one such example among many. The question that poses is: do we absolutely need the traditional state-form in the contemporary practices, talks, and approaches around biodiversity, such as, for example, the ecosystem approach? In any event, what are the legal and political tools for the representation of *interests of nature* beyond both the economic valuation of biodiversity and the anthropocentric vision?

5.1 The Concept of Ecological Solidarity in Motion

In order to explore the limits inherent to the approach of ecosystem services, we need to appreciate and understand the stakes through the prism of the notion of socio-ecosystem. For doing so, we need to explore few pertinent concepts.

For human communities interlinked with a habitat through residence or everyday activities, it is quite important to enlighten consequences of their development activities on ecological functions of given ecosystems (Rees 1998). Therefore, it appears that humans need adequate conceptual tools to help them to take into account not only what we know now about ecosystems and habitat, but also, and indeed, especially, what we do not know yet. The well-known precautionary approach in environmental law provides an expedient basis for recognition of scientific uncertainty[19] in case of threats of serious or irreversible damage. This approach, however, does not offer an all-inclusive perspective or solution for the socio-ecosystem. A holistic view is necessary in order to at least: (1) induce compliance (e.g., maintain and/or develop social behaviour in the way compatible with and not harming ecosystems' productive capacities - this means a way that allows ecosystems and societies who are depending on ecological process to preserve their reproduction and resilience capacities); (2) prevent, or, if occurred, resolve, conflicts (e.g., prevent environmental, social and economic inequities; prevent and manage the conflicts between humans and nature, and within human societies, for example, conflicts for resources and space); and (3) socialise participants (e.g., apprehend the environmental complexity of given ecosystems; include the diversity of social and economic actors).

5.2 The Principle of Ecological Solidarity

In order to better grasp the complexity of habitat in human development strategies—and more generally, in human behaviour—we need to look at it not only from a biological perspective, but also from social and cultural perspectives. In this respect, law is a very specific means to express a social project. Ideally, it is based on ethical value and its function to introduce regulation in social relationships.

In the context of our study, Antonio Negri, in his theories of "the law of the common", gives one pertinent example of how such regulation of social relationships could be envisaged:

[19]See Principle 15, Rio Declaration on Environment and Development, 1992: In order to protect the environment, the precautionary approach shall be widely applied by States according to their capabilities. Where there are threats of serious or irreversible damage, lack of full scientific certainty shall not be used as a reason for postponing cost-effective measures to prevent environmental degradation.

> A[nother] definition of the common is the sociological-institutional one. The development, from civil society to forms of public organization, to a common conceived as a societal outcome or as the result of the tendency to form associations, is seen just as a product of an ongoing activity. (. . .) Considered in its most recent figures, the "institutional" common is defined (for example, in Luc Boltanski's scholarship) by the abandonment of sociologies that focus on vertical dimensions and on the opacity of the alienated consciousness of the actors, for the benefit of a sociology that insists on horizontal relationships (and, of course, on networks) and on "contextualized" actions of actors led by strategic reasons or moral imperatives.[20]

A pertinent concept related to such vision of law, value and commons, particularly used in French law is the *ecological solidarity*. This principle lays down the recognition of the interdependence between humans, ecosystems, and their physical (natural or built) environment, as well as a new vision of nature conservation. The concept emerged in connection with the preparation of the amendments of Act on the National Parks (*la loi n°60-708 du 22 juillet 1960 relative à la création des parcs nationaux*). It has been taken up by *la réforme des parcs nationaux Loi n°2006-436 du 14 avril 2006,* and eventually, by *la loi de reconquête de la biodiversité, de la nature et des paysages, 20 juillet 2016.*

In general, the principle provides a comprehensive grid of analysis of the actual and potential environmental risks of the decrease or loss of ecosystem services related to the production of an ecosystem. Namely, it focuses on the interface of physical, living and human activities in various spatial and temporal patterns (from local to global, from short to long term) (Mathevet 2012; Thomson et al. 2011).

The majority of other jurisdictions are barely familiar with the exact concept of ecological solidarity as it is laid down in French law. Other connected concepts exist, such as ecological integrity (Westra et al. 2013) or connectivity (Bonnin 2008, 2012; Fleury and Prévot-Julliard 2010; Mitchell et al. 2013). Ecological integrity is used in different international agreements[21] and in national laws such as in the 1972 US Clean Water Act[22]. Ecological integrity tends to preserve ecosystems undiminished ability to continue "their natural path of evolution, their normal transition over time and their successional recovery from perturbation" (Westra 2005).

What is particular about the ecological solidarity is that it seeks to take into account the complexity of the socio–ecosystems, and thus can cover the following: (1) territorial solidarity; (2) social solidarity; (3) interconnexion between human and nature (The community of destiny, Mathevet 2012).

---

[20]Negri (2010), p. 21.

[21]The most famous one is the Great Lakes Water Quality Agreement (GLWQA), an agreement between Canada and the United States, first signed in 1972. It contributes to restore and protect the chemical, physical and biological integrity of the waters of the Great Lakes (article 1).

[22]The Clean Water Act (CWA) establishes the basic structure for regulating discharges of pollutants into the waters of the United States and regulating quality standards for surface waters. The basis of the CWA was enacted in 1948 and was called the Federal Water Pollution Control Act, but the Act was significantly reorganized and expanded in 1972.

The so-called 2016 French 'biodiversity law' generally provides for a set of tools and mechanisms, such as the creation of the French Agency for biodiversity, the implementation of the Nagoya Protocol, a compensation of environmental damages, the strengthening of the compensation mechanisms, new legal tools of protection and preservation of nature, and tougher sanctions. In particular, Title I, Art.2 of the bill ("Fundamental principles") contains the principle of ecological solidarity (together with the principle of the triple "avoid, reduce, compensate" and the very notion of the biodiversity): namely, the principle calls *to take into account, in public decision-making that may have a significant impact on the environment of the concerned territories, the interdependence between humans, ecosystems, and their physical (natural or built) environment.*[23] It results from this definition that mainly one specific aspect of ecological integrity is designed in this new legal principle: that is, the territorial solidarity. It also re-defines the responsibilities of humans, bringing them away from using and protecting distinct species and spaces towards a more holistic approach to the impacts of human activities on any forms of life.

With regards the questions of representation raised earlier in this paper, it is interesting to note that it is the *region* (and not the state) that is designated in the bill as "the steering point" ("*l'échelon de pilotage*").[24] Its efficient integration into space planning and conservation management strategies will require, however, the *collective* use and management by local *communities* and stakeholders of the diverse facets of ecological solidarity. This could hopefully shed light on future discussions on the role of representation, and standing of local communities, territorial collectivises, and on the so-called horizontal subsidiarity in the considerations regarding the commons.

Being quite out of the scope of this analysis, it is still crucial to briefly sketch numerous aspects of the 2016 French biodiversity law that seem to bring it 'ahead to the past' of economic valuation of nature and biodiversity. For example, in and through its mechanisms of compensation, the bill's objective is to evaluate and tackle the infamous 'net' losses that we addressed earlier in this analysis, hence bluntly taking the ecological, not economical, value of biodiversity, as a primary accounting unit. Even worse, the 'efficiency' of concrete compensation measures to take is conditioned by the obligation to demonstrate concrete *results*. One may only wonder how humans envisage showing concrete results of substituting biodiversity by some sort of compensation. The bill also conditions and justifies biodiversity conservation by the benefits from the services it brings to humans. In its Article 2, the bill says that the environment represents a heritage (*patrimoine*) of the nation and generates "ecosystem services and use values"[25]. In this way, French legislator seems to pave the groundwork for the opportunities to valorise nature through

[23]http://www.senat.fr/dossier-legislatif/pjl14-359.html.

[24]Id.

[25]Complete article 110-1. I of French Environmental Code.

markets, thus reinforcing utilitarian approach to nature based on mainstream economic view of use of ecosystems, discussed in Sects. 2 and 3 of our analysis.

All in all, resulting from an uneasy political compromise, the above French law tries to strike a balance between mainstream political pressure to recognise the ecosystem services and the necessity to go further in public policies concerning ecosystems preservation in relation with spatial management and ecological connectivity.

5.3 Towards Relational Justice

Notably, as the latest French bill discussed just above is about *natural parks*, it is worth reminding that, historically, most national parks, and similar protected areas have been established for *humans*, everywhere in the world, and not just in postcolonial nations (e.g. Barrère et al. 2005; Doak et al. 2015). What Wilson (2016a, b) proposes is that these parks need to be created for nonhumans as well, evoking inter-species equity.[26] While the term 'environmental justice' often refers to (un)equal distribution of environmental burdens and benefits across human populations (e.g. Low and Gleeson 1998), the concepts such as 'inter-species equity' or 'biospheric egalitarianism' refer to integrity of ecosystems, and hence seem closer to justice between all species (Wissenburg 1993; Baxter 2005; Schlosberg 2007; Cafaro and Primack 2014; Kopnina 2016; Cafaro 2015).[27]

Overall, if we would like to use the principle of ecological solidarity, drawing on lessons learnt from the original 1960 French policy, we should value the *natural spatial and temporal interdependence* among entire ecosystems. This sets the scene for a new vision of nature conservation, which is intrinsically connected with the ecosystems approach. In our vision, this approach being about the protection and conservation of entire ecosystems rather than distinct species or sectors, constitutes the entry level for implementing the ecological solidarity. The latter thus offers a way between, but also beyond *ecocentric* and *anthropocentric* ethics. Namely, to go beyond, we should at least depart from the sovereign models of central and buffer zones and set protected areas within their wider socio-ecological contexts. Therefore, the principle of ecological solidarity implies responsibility regarding not (only) current and future generations of humans, but generations of entire ecosystems (the latter including humans), as well as regarding current and future biodiversity.

---

[26]All in Kopnina, pp. 181.

[27]All in Kopnina, p. 181-2.

6 Concluding Remarks

The principle of ecological solidarity promotes an innovative approach for the recognition of the complexity of ecosystems in general. While still being recognised in French law in quite a restrictive way, however, the principle could find many other applications, including at international level, to overcome the traditional—both contestable and contested—binary vision "man-nature". Its application could lead to the revision of the traditional instruments of ecosystem protection and management.

References

Apostolidis C, Chemillier-Gendreau M, Pontier JM (1993) L'apport de la dialectique à la construction de l'objet juridique. Contractualisation et planification. Revue du Droit Public 3:611–694

Barrère C, Barthélémy D, Nieddu M, Vivien F-D (éds) (2005) Réinventer le patrimoine. De la culture à l'économie, une nouvelle pensée du patrimoine ?, Collection Gestion de la culture, L'Harmattan

Barritt E, Viñuales JE (2016) Legal scan: a conservation Agenda for biodiversity beyond national jurisdiction. Cambridge centre for environment, energy and natural resource governance. University of Cambridge, Cambridge

Baxter BH (2005) A theory of ecological justice. Routledge, London/New York

Bonnin M (2012) L'émergence des services environnementaux dans le droit international de l'environnement : une terminologie confuse. VertigO, Vol. 12 n° 3, décembre, en ligne

Bonnin M (2008) Prospective juridique sur la connectivité écologique. Revue Juridique de l'Environnement, pp 167–178

Cafaro P (2015) Three ways to think about the sixth mass extinction. Biol Conserv 192:387–393

Cafaro P, Primack R (2014) Species extinction is a great moral wrong. Biol Conserv 170:1–2

Callon M (dit.) (1998) The laws of the markets. Blackwell, Oxford

Carpenter SR, DeFries R, Dietz T, Mooney HA, Polasky S, Reid WV, Scholes RJ (2006) Millennium ecosystem assessment: research needs. Science 314:257–258

Churchill R (2015) The LOSC regime for the protection of the marine environment – fit for the twenty-first century? In: Rosemary R (ed) Research handbook on international marine environmental law (Research handbooks in environmental law). Edward Elgar, Cheltenham

Costanza R, D'Arge R, De Groot R, Farber S, Grasso M, Hannon B, Limburg K, Naeem S, O'Neill RV, Paruelo J, Raskin RG, Sutton P, Van Den Belt e M (1997) The value of the world's ecosystem services and natural capital. Nature 387:253–260

Costanza R, Daly HE (1992) Natural capital and sustainable development. Conserv Biol 6:37–46

Chevassus-au-Louis B, Salles J-M, Pujol J-L, Bielsa S, Martin G, Richard D (2009) Evaluation économique de la biodiversité et des services liés aux écosystèmes : contribution à la décision publique. Rapport du CAS, Paris

Daly HE (1997) Beyond growth: the economics of sustainable development. Beacon Press, Boston

Doak DF, Bakker VJ, Goldstein BE, Hale B (2015) What is the future of conservation? In: Wuerthner G, Crist E, Butler T (eds) Protecting the wild: parks and wilderness, the foundation for conservation. The Island Press, Washington/London, pp 27–35

Doussan I (2009) Les services écologiques: un nouveau concept pour le droit de l'environnement? In: La responsabilité environnementale, prévention, imputation, réparation, C. Cans (dir.), préface de G.Viney, Dalloz, Coll. Thèmes et commentaires, 421p, pp 125–141

De Groot RS (1987) Environmental functions as a unifying concept for ecology and economics. Environmentalist 7(2):105–109
Fitoussi J-P., Laurent E (2008) La nouvelle écologie politique, Seuil et la République des idées
Fleury C, Prévot-Julliard A-C (2010) L'exigence de la réconciliation. Biodiversité et Société, Fayard
Froger G, Méral P, Le Coq J-F, Aznar O, Boisvert V, Caron A et Antona M (2012) Regards croisés de l'économie sur les services écosystémiques et environnementaux. VertigO 12(3), décembre, en ligne
Gómez-Bagghetun E, Groot R, Lomas P, Montes e C (2010) The history of ecosystem services in economic theory and practice: From early notions to markets and payment schemes. Ecol Econ 69:1209–1218
Hegel GFW (1977) Phenomenology of spirit (trans: Miller AV). Oxford University Press, Oxford
Hervé-Fournereau N, Langlais A (2013) Chapter 3: does the concept of ecosystem services promote synergies between European strategies for climate change and biodiversity? In: Biodiversity and climate change linkages at international, national and local levels the IUCN academy of environmental law series. Edward Elgar, pp 65–93
Hautereau-Boutonnet M (2015) Le contrat et l'environnement en droit comparé. Bruylant, Bruxelles
Hoffmeister J (ed) (1977) Force and the understanding: appearance and the supersensible world: phenomenology of spirit, 5th edn. Oxford University Press, New York
Jansson AM, Hammer M, Folke C, Costanza R (eds) (1994) Investing in natural capital: the ecological economics approach to sustainability. Island Press, Washington
Kopnina H (2016) Half the earth for people (or more)? Addressing ethical questions in conservation. Biol Conserv 203:176
Kourislky P, Geneviève VG (1999) Le principe de précaution, Rapport au Premier ministre, Édition Odile Jacob, Paris.
Low N, Gleeson B (1998) Justice, society and nature. Routledge, New York
Maris V (2014) Nature à vendre. Les limites des services écosystémiques. Quae, Versailles
Maris V (2016) Philosophie de la biodiversité. Petite éthique pour une nature en péril, 2nd edn. Buchet/Chastel
McClintock A (1995) Imperial leather: race, gender, and sexuality in the colonial context. Routledge, New York
MEA (2005) Ecosystems and human well-being: a framework for assessment, millennium ecosystem assessment. Island Press, Washington
Mitchell MGE, Bennett EM, Gonzalez A (2013) Linking landscape connectivity and ecosystem service provision: current knowledge and research gaps. Ecosystems 16:894–908
Morgera E, Tsioumani E, Buck M (2014) Unraveling the Nagoya Protocol: a commentary on the Nagoya Protocol on access and benefit-sharing to the convention on biological diversity. Brill nv, Leiden
Negri A (2010) The law of the common. In: Finnish Yearbook of International Law (Vol. 21)
Naim-Gesbert E (2014) Droit général de l'environnement, 2e edn. Lexis Nexis, Paris
Prevost B, Rivaud A, Michelot A (2016) Économie politique des services écosystémiques : de l'analyse économique aux évolutions juridiques. Revue de la régulation [En ligne], 19 | 2016, on line 31 août 2016, URL: http://regulation.revues.org/11848
Prieur M, Bétaille J, Cohendet M-A, Delzangles H, Makowiak J, Steichen P (2016) Le droit de l'environnement. Dalloz, Paris
Rees W (1998) How should a parasite value its host? Ecol Econ 25:49–52
Roche P, Geijzendorffer L, Maris V (coord.) (2016) Valeurs de la biodiversité et services écosystémiques. Perspectives interdisciplinaires. Quae, Versailles
Schlosberg D (2007) Defining environmental justice: theories, movements, and nature. Oxford University Press, New York
Sikor T (2013) The justices and injustices of ecosystem services. Routledge, New York, Canada

Strang V (2016) Justice for All: Inconvenient truths and reconciliation in human-nonhuman relations. In: Kopnina H, Shoreman-Ouimet E (eds) Routledge handbook of environmental anthropology. Routledge, New York, pp 263–278

TEEB (2010) The economics of ecosystems and biodiversity ecological and economic foundations. Edited by Pushpam Kumar. Earthscan, London

Thompson JD, Mathevet R, Delanoe O., Gil-Fourrier C, Bonnin M., Cheylan M (2011) Ecological solidarity as a conceptual tool for rethinking ecological and social interdependence in conservation policy for protected areas and their surrounding landscape, C.R Biologies. Elsevier

Westra L (2005) Ecological integrity. In: Mitcham C (ed) Encyclopedia of science, techonoly and ethics, vol 2. Macmillan, Detroit, pp 574–578

Westra L, Taylor P, Michelot A (eds) (2013) Confronting ecological and economic collapse. Ecological integrity for law, policy and human rights. Routledge, London

Wilson EO (2016a) Half earth. Liveright, New York

Wilson EO (2016b) Half earth. https://aeon.co/essays/half-of-the-earth-must-bepreserved-for-nature-conservation

Wissenburg M (1993) The idea of nature and the nature of distributive justice. In: Dobson A, Lucardie P (eds) The politics of nature: explorations in green political theory. Routledge, London

Democracy, Sovereignty and the Challenge of the Global Commons

Klaus Bosselmann

This chapter looks at forms of governance that are reflective of the need for protecting the integrity of ecological systems. The existing system of global environmental governance has been stunningly inadequate in stemming the flow of environmental damage. Despite more than 900 environmental treaties in the past 40 years,[1] human induced environmental damage is reaching levels where the International Council for Science (ICSU) warns that we have reached a point where the continued functioning of the earth system as we know it, is at risk[2]:

> This science also tells us that the rate of global environmental change is, so far, vastly outpacing our response and thus, our current path is unsustainable. We know enough to state with a high degree of scientific confidence that without action to mitigate drivers of dangerous global change and enhance societal resilience, humanity has reached a point in history at which changes ... will undermine development prospects and cause significant human suffering associated with hunger, disease, migration and poverty. If unchecked or unmitigated, these changes will retard or reverse progress towards broadly shared economic, social, environmental and developmental goals.

For assistance with researching and writing this chapter I am indebted to Kristen Jones, New Zealand Centre for Environmental Law; for the wider context of this chapter see *Creating a Safe Operating Space through Law and Governance* (146 pages, Sept. 2016), report prepared by Klaus Bosselmann and Kristen Jones for the Planetary Integrity Project http://planetaryboundariesinitiative.org/.

[1]Frank Biermann and Steven Bernstein "How Rio+20 Can Herald a Constitutional Moment" (15 March 2012) The Guardian www.theguardian.com. See also International Geosphere-Biosphere Programme (IGBP) "Governance experts warn UN overhaul required to govern earth system" (press release) 23 November 2011 www.igbp.net.

[2]International Council for Science (ICSU) (2010), p. 5.

K. Bosselmann (✉)
New Zealand Centre for Environmental Law, University of Auckland, Auckland, New Zealand
e-mail: k.bosselmann@auckland.ac.nz

L. Westra et al. (eds.), *The Role of Integrity in the Governance of the Commons*,
DOI 10.1007/978-3-319-54392-5_4

Any real change will require a move away from a 'state-centered' global governance system where acceptance of and then compliance with treaties is voluntary, to an 'earth-centered' system in which states act as trustees of the environment on behalf of all citizens. However, nothing short of a renewed push by civil society based on strong ethical values will achieve this. Citizens everywhere must see themselves as stewards of the earth. This vital paradigm shift in how we view 'our home' and think about the environment is the only driver strong enough to create the right institutions, needed at local, regional and global levels to reverse environmental destruction.

What are some of the hurdles we face in creating these institutions? At our GEIG meetings, we have regularly discussed the paradox of seemingly unstoppable economic growth in a closed planetary system with ecological boundaries. There are further paradoxes embedded in the complexity of the Anthropocene that are also clear barriers to a new governance system.

1 The Democracy Paradox

The latter part of the twentieth century and the early twenty-first century have seen a dramatic expansion of democracy as a governing concept.[3] 120 sovereign states out of 192 contain democratic political systems and encompass 58% of the world's population.[4] Patrick Deneen describes this global move towards democracy, as "the only regime most living humans now deem worthy of serious consideration".[5] Democracy generally defined is "a system of government by the whole population" typically through elected representatives.[6] In a nutshell, the core idea behind democracy is that ultimate authority and power remains with the people. It is a powerful philosophy focused on the innate freedom of every person 'to be left alone' or to function as they please. The primary role of any democratically elected government is to ensure the protection of this freedom not only for its citizens but also for the state itself in relation to other states.[7] Yet the paradox hidden within democracy in the Anthropocene is that, as Eyal Benvenisti describes it[8]:

[3]Bosselmann (2015), p. 9.

[4]Ibid. See also, Archibugi (2002), p. 27.

[5]Deneen (2005), p. xvi.

[6]Definition of 'democracy' Oxford dictionaries, www.oxforddictionaries.com.

[7]*Creating a Safe Operating Space through Law and Governance* (146 pages, Sept. 2016) at 26. Democracy was espoused and popularized by John Locke, Jean-Jacques Rousseau and Immanuel Kant.

[8]Eyal Benvenisti "The future of sovereignty: The nation state in the global governance space" (2015) 01/2015, The Global Trust Working Paper Series www.globaltrust.tau.ac.il/publications.

> In our era of global governance the freedom to be left alone no longer holds the promise of providing citizens with control over their lives because no Chinese walls will be capable of insulating communities from the outside. States that seek to ensure freedom to their citizens must act proactively by engaging foreign and international governance bodies, and by ensuring opportunities for their citizens to do the same.

Globalization and technology mean we are more interconnected than ever before. Economic activity and trade flows through porous national borders creating state interdependency. Open trade brings more prosperity for countries but also worsens ecological footprints as we crisscross the oceans using them as highways to move goods from one side of the world to the other. The democratic vacuum at the global level has invited powerful states to hollow out single country agendas in the international space,[9] to negatively effect weaker states and to allow large multinational corporations (with private profit agendas) to hijack economic rules.[10] This development has eroded state autonomy, reducing democracy through privately negotiated trade deals taking place outside the public democratic arena and reducing states' ability to regulate their own affairs. Of particular concern is the flow on effects which compromise states' internal social and environmental agendas. States see it as their duty to participate in this economic model in order to secure prosperity for its people—if you don't have a seat at the table, your family will go hungry[11]:

> Needs of the market and needs of society are not the same and must be carefully separated. The fact that governments have increasingly neglected this separation, points to a crisis of democracy. Quite obviously, it is not in the interest of the 'demos' if governments allow markets to determine the direction, scope and content of public policies. Democratic process "requires that citizens decide all issues that are politically decidable"

Democracy now requires deeper construction at the global level if our freedoms are to remain, along with the realization that with any freedoms also come duties and responsibilities. The Anthropocene traffics in complex global environmental problems, which no single state can address like climate change, ocean acidity and acid rain and which have no respect for national borders. This changes the playing field of democracy from an activity that happens only within state borders to include governance systems, activities and institutions that are global. Immanuel Kant foresaw this development where international society is forced to move slowly and in a non-linear fashion towards global democracy.[12] Benvenisti describes this leavening process taking place (and not particularly desired by states) as "engaged sovereignty"—the growing recognition by democratic states of their embeddedness within a global order to which it is accountable. Kant puts it this way[13]:

---

[9]Bosselmann (2015), p. 12.

[10]Ibid. at 10, 11,12, 13, 18, 30, 31, 38 and 45.

[11]Ibid. at 12.

[12]Ibid. at 11.

[13]Immanuel Kant "Idea for a Universal History from a Cosmopolitan Point of View" Seventh Thesis (1784).

> The same sociability which drives man to the creation of a commonwealth causes any single commonwealth to stand in unrestricted freedom in relation to others; consequently, each of them must expect from another precisely the evil which oppress the individuals and forced them to enter into a lawful civic state. The friction among men, the inevitable antagonism, which is a mark of even the largest societies and political bodies, is used by nature as a means to establish a condition of quiet and security.

The process of global democracy is far from complete but is at the root of unlocking solutions to the ecological crisis. The role of the citizen is expanding from the national to the international scale with citizens all over the world seriously concerned about the environmental crisis and the ineffectiveness of the existing system to address the problems.

To date, the strongest expression of civil society that announces a new ethic to take responsibility for the earth is found in the Earth Charter.[14] This provides a new framework for global governance and a new 'covenant' between humans and nature.[15] Part IV of the Charter sets out four principles for sustainable earth governance, which are[16]:

1. To strengthen democracy at all levels (including transparency, accountability, participation and access to justice);
2. To integrate the required capabilities for a sustainable way of living in a process of lifelong learning;
3. To treat all living beings with respect and consideration;
4. To promote a culture of tolerance, nonviolence, and peace.

The Earth Charter is not sanctioned by sovereign states but by global citizens[17]:

> A multilayered transnational system of democratic governance based on constitutional principles such as those expressed in the Earth Charter may still involve states, but is not dependent on them. . .. Any democratic system is accountable only to the demos, neither to states nor corporations nor any other legal constructs.

Despite decades of political discussion, only one treaty has set targets and timetables (the Kyoto Protocol) to take practical action on just one serious threat—climate change, with no practical success so far. This reflects the failure of the current international approach of negotiated consensus as a viable method to contain and reverse global environmental problems. We need to acknowledge the existing system of governance is not working and as citizens, explore and demand a dramatically improved system. This calls for re-imagining and creating institutions at the local, regional and global levels that gain legitimacy through the process of global democracy and have the necessary "teeth" to take the required actions. How do we honor the philosophical heart of the idea of democracy by placing decisions back in the hands of the public? How do we ensure people most affected by

---

[14]The Earth Charter Initiative, *The Earth Charter* (2000) wwwearthcharterinaction.org.

[15]Bosselmann (2015), pp. 39–47.

[16]Ibid. at 2. See also Engel (2010), pp. 29–40.

[17]Bosselmann (2015), p. 3.

ecological damage have a democratic voice? Essentially this involves a shift away from state centered governance to earth focused governance emphasizing the role of the citizen as the source of legitimacy. Re-conceptualizing and re-invigorating state sovereignty is a fundamental part of this process.

2 The Sovereignty Paradox

In the current system of law and governance, the earth is conceptually divided into two distinct areas. The first is that of sovereign states and the second is the 'global commons' which consists of everything outside (or not controllable) by states such as the atmosphere, the oceans, seabed and fisheries (beyond coastal state limits), Antarctica and outer space (including celestial bodies), the geo-stationary orbit and the electromagnetic spectrum.[18] Westphalian sovereignty and international environment law provide that:

1. Sovereign states effectively 'own' the natural resources and the environment within their borders and may preserve, use or exploit that environment as they see fit and without external interference.
2. With respect to that part of the environment *other* nation states 'own' the international principle of good neighborliness (not to cause damage to another state) applies.
3. With respect to the global commons, these are not owned by anyone and access to them is free to all. This results in the 'tragedy of the commons'[19] where a natural resource becomes overused and eventually exhausted because no limits on use are set.

   Nature itself has no legal rights or 'standing' in the eyes of the law (no legal personality) therefore any overuse, harm or pollution that occurs whether by a state, a corporation or an individual is difficult to rectify through normal legal channels. There is no legal entity to bring an action against, no property right breached and no plaintiff.

   There are a multitude of environmental treaties and soft law documents for discrete parts of the global commons, which have been developed over the last 40 years through international negotiation amongst states and with the encouragement of the United Nations and civil society. However sovereign states *choose* whether to be bound by these treaties or not and most of them operate with minimal success in terms of protecting the environment. Each treaty operates independently with legal autonomy, which results in problem shifting,[20] exponential complexity of procedures and rules and a growing

---

[18]Ibid. at 72.

[19]Hardin (1968), pp. 1243–1248.

[20]Problem shifting occurs where one treaty ignores environmental problems outside its own narrow mandate. For example the UNFCCC treats carbon capture by the oceans as a carbon

administrative burden on states. Where there are 'gaps' in environmental law that are not covered by a either a treaty regime at the international level or legislation at the national level, the vacuum is filled by default with either sovereign state legal rights or individual property rights that allows exploitation of the commons without accompanying duties or responsibilities to do so in a sustainable manner.[21]

International environmental law is based on respect for state sovereignty, which is the international mirror of private property rights.[22] However the paradox within state sovereignty is that an examination of its history uncovers the potential for understanding state sovereignty as a trustee for the preservation and protection of the ecological integrity of the environment. The United Nations has a history of trusteeship concepts that we can draw upon, which itself is based upon sovereign state trusteeship traditions.[23]

3 Reclaiming Earth for Global Citizens

We have already briefly discussed the notion that political power and the legitimacy of a sovereign state flows from representing its people[24]:

> Popular sovereignty denotes that the state's sovereign powers belong to the people, and so those powers are held in trust by their rulers on condition that they be used for the people's benefits. Popular sovereignty thus implies that the state and its institutions are fiduciaries of the people; for the justification rests exclusively on the authority they enjoy governing and serving the people.

The culture of the Anthropocene has birthed a new and more cosmopolitan citizen who is often more mobile, moving and living in different countries for work, and connected via technology with others around the world. Even those citizens living in less developed countries share this connectivity and are aware of matters far outside of their immediate environment. The idea of 'citizenship' is no longer confined to nations. In fact, if we identify only as a citizen of one country we undermine the very global conditions we need to flourish.[25] If we reframe our thinking of 'home' from our immediate country to the earth itself, and envisage that we are a part of the earth system and the 'community of life' relying upon a shared

---

sink rather than acknowledging the environmental impact of excess CO_2 on the oceans. See Kim (2013), p. 89.

[21]Jóhannsdóttir (2009).

[22]Bosselmann (2015), p. 167.

[23]Bosselmann (2015), pp. 198–232.

[24]Criddle and Fox-Decent (2009), p. 350.

[25]Bosselmann (2015), p. 31.

healthy biosphere, this brings an awareness of the need for institutions that reflect this reality.[26]

> The difference between humans and nature is not absolute or categorical, but gradual. This aesthetic experience offers the moral insight of a special responsibility for the nature we are essentially part of.

Once we approach our relationship with the environment in this paradigm fracturing fashion, an interesting question naturally arises. Who actually owns the earth? Is it us? What happens if the public asserts property rights over the natural environment? What if there is an insistence that environmental public goods are owned by all of us? As legal owners, the public could charge for damage to 'common property' and provide rewards to those who protect it. Private property rights and state sovereignty would continue to exist—but only up to the place where common property rights begin.[27] This idea to assert global citizen rights to the environment within the legal system is what Donella Meadows would call a key 'leverage point'—meaning, a point of power within a complex system where a small shift can produce big changes.[28] Such a change of mindset or paradigm is only possible through a radical shift in our relationship with the earth, moving away from viewing earth as just a resource to create economic growth towards seeing it as a birthright to be protected by and for ourselves, for future generations and for all other forms of life. The driver to any change towards 'responsible sovereignty' rests with real people embracing the idea that democratic citizenship operates at local, national and global levels. If we begin to see ourselves as protectors and stewards of the earth and our sovereign states as trustees of the earth for the common good on our behalf, this is the *crucial step* that takes us towards a sustainable system of governance.[29] Meadows points out that the striking aspect of living and social systems is their ability to change themselves by creating new paradigms, structures and behaviors[30]:

> In biological systems that power is called evolution. In human society it's called technical advance or social revolution. In systems lingo, it's called self organization… a system that can evolve, can survive almost any change, by changing itself… the same could be said of human cultures of course, which are the store of behavioral repertoires, accumulated over not billions, but hundreds of thousands of years. They are a stock out of which social evolution can arise.

Paradigms are the source of our systems. From them flow our social agreements about reality along with system goals, information flows, and feedbacks.[31] Thomas

---

[26]Ibid. at 37.

[27]Bosselmann (2016), pp. 270 and 271.

[28]Meadows (1999).

[29]Bosselmann (2015), p. 289.

[30]Bosselmann (2016), pp. 15 and 16.

[31]Ibid. at 18.

Kuhn wrote the seminal book about paradigm shifts but in Meadows words, the way to create such a shift involves the following[32]:

> In a nutshell, you keep pointing at the anomalies and failures in the old paradigms, you keep speaking louder and with assurance from the new one; you insert people with the new paradigms in places of public visibility and power. You don't waste time with reactionaries; rather you work with active change agents and with the vast middle ground of people who are open-minded.

Having thrown down the gauntlet to take charge of human destiny, we need new institutions that reflect and operationalize this.

4 The State Trusteeship Concept for the Global Commons

The global commons as we normally think of them are those areas outside of states jurisdiction. However a broader understanding also includes things existing across, i.e. within and outside, state boundaries like water, the atmosphere and biodiversity. This broader understanding of the commons provides framework we need to appreciate 'ownership' of the commons. This kind of ownership is with all humanity, includes communal possession and is fundamentally different from 'property' that might be exclusive and could be sold or commercialized. Ownership is understood here not as ownership in our individual capacity but as the human community reflecting the idea of the common heritage of life.[33] This principle properly understood, creates humanity and all other life as the deserving new 'subjects' of international law and expands sovereign state responsibility beyond territorial borders. Collective state trusteeship can be described as protecting all areas of the global commons, inside and outside of state borders that are recognized as the common heritage of life.[34]

For many years the United Nations has considered the question of the duty of states to act as trustees for the global commons.[35] Currently the state sees its role largely as a facilitator of economic growth[36] and a protector of private property.

[32]Ibid.

[33]The common heritage of mankind (CHM) and the common heritage of life (CHL) doctrines, see above n 4 at 75–79, 108–111 and 248. See also Weiss (2012), p. 153, and Taylor (2016), pp. 104–130.

[34]The ideas discussed here go further than the 'common concern of mankind'(CHM) principle by arguing that the 'common heritage of life' (CHL) prevents appropriation of 'common goods' for individual interests; the CHL concept was developed by Taylor (1998). See also above n 4 at 248 and Redgwell (2005), p. 179.

[35]Bosselmann (2015), pp. 233–244.

[36]A recent example of this at the local government level in New Zealand, is a case where the Ashburton District Council was selling a piece of land in its business estate to an overseas bottled water company with a resource consent to extract up to 45 liters of water per second from the towns aquifers. Local residents campaigned against the sale but District Mayor, Angus McKay

Governments must function within short-term political cycles. As a result the system is structured to encourage short-term solutions. However, environmental trusteeship of states need not be seen as standing in opposition to state sovereignty. An insistence that the global commons belong to all, now and into the future, requires an understanding that governance has the common good as its focus. Maintaining a healthy environment is essential for human prosperity. As representatives of the governed, states can ensure its people ecological, social and democratic well being. Attention shifts from a negotiated consensus settling for the lowest common denominator amongst states (and leaving the commons unprotected) towards a focus on trusteeship governance of the earth as the best approach to foster non-competitive cooperation towards managing the commons.

The public trust doctrine is a simple mechanism that can support the idea that the global commons should be held in trust as assets to serve the public good.[37] The responsibility of sovereign states would be to protect these assets from harm and ensure their use for the public and future generations. Nationally a government would act as environmental trustee and internationally, states would act jointly as trustees of the global commons.[38] The idea of global trusts has been promoted by Peter Sand, Mary Wood, and Peter Barnes among others.[39] We have examples within the United Nations of global institutions that have a trustee mandate. These include the (historical) UN Trustee Council which oversaw the transition of territories from colonialization to independence, the World Health Organisation (WHO) holding governance powers with respect to public health and the World Trade Organisation (WTO) in relation to trade matters.[40]

5 What Would a Trust for the Global Commons Look Like?

1. A legal entity or trustee such as a World Environmental Organization (WEO) would be required that is responsible for the protection of the beneficiaries, being present and future generations and the community of life. The trustee would exercise discretion and the administration of the trust corpus (the biosphere) in the beneficiaries' interest. Because the corpus is common to all, the commons are insulated from individual claims by companies, governments or specific individuals.[41]

"said the Council had the public's best interests at heart, as the sale of the site would go towards rates remissions". See Sally Murphy "Council to Sell Rights to Bottled Water Company" (4 April 2016) Radio New Zealand www.radionz.co.nz.

[37]Sand (2013). See also, Sax (1969), p. 471; Redgwell (1999), Coghill et al. (2014).

[38]Bosselmann (2016), p. 271.

[39]At 267–280.

[40]Bosselmann (2015), pp. 199–232.

[41]Ibid. at 152.

2. Current generations would be the donors or settlers (as well as the beneficiaries) of the trust.
3. Beneficiaries would consist of humans and other life, which currently have no existing legal personality. In other words the beneficiaries are *the community of life*. This gives nature standing which is an important new aspect of any proposed trusteeship regime and which reduces the likelihood of environmental governance simply being about preventing the worst harm and slowing exploitation as has happened so far. Such a trust concept approaches the trusteeship obligations from the perspective of first restoring and then maintaining ecological integrity of the biosphere. This is its focus. Unless a WEO (or similar institution) has equivalent powers to the existing WHO and WTO failure is a distinct possibility. Likewise the necessary funding uncontaminated by political pressures is integral to success.[42]
4. The exercise of the trustee's discretion would be framed by a number of fundamental obligations or guiding principles which are:

   - Sustainability.
   - The common heritage of life doctrine, giving nature standing alongside humans.
   - Decisions to be guided by the precautionary principle.
   - Intergenerational equity, ensuring that the planet is passed on to future generations in a condition capable of sustaining life.
   - Intragenerational equity, providing a legal voice to the less powerful in the global community both human and nonhuman. This provides the trustees with a significant role in managing the equitable use of the global commons amongst developed and developing nations.[43] Serious thought about voting rights and structure that would allay developing countries concerns around power ratios would require careful consideration.[44]
   - The ecological integrity of the biosphere perhaps through recognition of a legal grundnorm.

6 Advantages of a World Environmental Organization

A WEO would have a number of advantages. Political friction between states (caused by domestic interests and competitive economic pressure) would be reduced in the face of a clear trusteeship regime applying equally to all. As long as states compete for natural resources, the negative feedback loop race towards the

---

[42]Ibid. at 264. The experience of the UNEP has highlighted the problems faced by underfunding.
[43]Ibid. at 153.
[44]Biermann (2011), p. 10.

bottom to use up resources continues.[45] Collective action must be *at least* at the same scale as collective harm.[46]

> The world community would benefit from the presence of an authoritative environmental voice in the international arena and a recognized forum for national officials and other stakeholders to work cooperatively to address global issues.

Currently earth system governance is not accountable to any one international institution solely devoted to supporting the biosphere.[47] The existing institutional structure is highly fragmented and spread across many institutions (too many cooks in the kitchen) with little coordination resulting in duplication of effort, funding, personnel and lack of strategic planning.[48] At the time the UN Charter was implemented in 1945, international security not the environment was the prime concern. The term 'environment' does not even appear in the UN Charter.[49] It wasn't until the early 1970s that the UNEP was set up,[50] "as a mere program, without legal personality, without budgets, and—according to its founding instruments, only with a 'small secretariat'". Frank Biermann points out that the UNEP is no match to other specialized international organisations and has had to compete for "time and attention and resources with more than a dozen other UN bodies".[51]

> The existing international environmental institutions, especially UNEP, are hampered by narrow or vague mandates, small budgets, and limited political support. No one organization has the political authority, vitality, expertise, and profile to serve as the center of gravity for the international environmental regime and to exert sustained political influence and other global fora

A WEO would provide high-level expertise and accountability along with a long-term orientation more suited to a global institution then to state governments. Core functions would be, decision-making, implementation, monitoring and dispute resolution. Esty and Ivanoa point out that sound environmental decision making hinges on some necessary factors. First, scientific assessment of high quality information to enable long-term forecasting of environmental trends. Second, the ability to approach problems from multiple perspectives and to draw on a wide range of information sources. Third, the establishment of guidelines and international norms, which over time could develop into more formal rules.[52]

---

[45]Esty and Ivanoa (2001), p. 4.

[46]Ibid. at 5.

[47]Biermann (2011), p. 5.

[48]For example UNEP, UNDP WMO and the OECD all have climate change programs with little strategic coordination. See above n 46 at 7.

[49]Biermann (2011), p. 6.

[50]The UNEP was setup in 1973 following the 1972 Stockholm Conference on the Human Environment and is not an intergovernmental institution but a subsidiary body of the General Assembly.

[51]Esty and Ivanoa (2001), pp. 7 and 8.

[52]Ibid. 10 and 11.

States authorisation of a WEO would lend legitimacy to any enforcement measures. A judicial arm similar to the dispute settlement body of the World Trade Organisation would encourage the fleshing out and development of the guiding principles suggested above. Environmental interests do not currently fit well into the law. Further administrative and judicial remedies are needed.[53] The growing area of international environmental administrative law regarding procedural rights is a possible area of evolution. Trustees of a WEO would include the sovereign states and importantly, also other parts of civil society.[54] Earth governance then becomes a much broader affair.

A WEO could be created by treaty, or by General Assembly Resolution that would endow it with independent institutional status and the power to resolve disputes. Christopher Stone has proposed the creation of a global commons trust fund[55] able to impose levies on use of the commons ensuring a dependable flow of money each year.[56] This mechanism would secure funding directly from those most responsible for using the commons.[57]

> Trusteeship says that the duty to protect the environment should not have to be negotiated. Sustainability is the goal - this will never change. Like justice and human rights, sustainability should be a foundational norm of humanity that guides the functions of states rather than be guided by them. Territorial sovereignty over natural resources... can only be perceived, therefore, as inherently linked to an obligation to protect the environment. The use of resources is covered by territorial sovereignty only within the parameters of ecological sustainability. This would read as: 'the sovereign right of states to exploit their resources pursuant to the principle of sustainability.'

7 Trusteeship Within States

So far, we have been talking about trusteeship at the global level but the idea of trusteeship and acting as guardians of the environment is one that needs to permeate through all levels of society to be effective. In New Zealand, a first experiment of 'rights for nature' can be found in a recent deed of settlement signed between the Whanganui River's Iwi (the local indigenous people) and the New Zealand government. The deed of settlement recognizes the river itself [58] as a 'legal person' with standing before the courts. Two human trustees will be appointed, one by the

[53]Bosselmann (2015), pp. 265–267.

[54]For example, the IUCN, the World Wildlife Fund, Greenpeace and the World Meteorological Organisation.

[55]Bosselmann (2015), p. 265. See also, Stone (1993), p. 34.

[56]Governments are well aware that institutions can only function with adequate funding. See above n 1 at 264.

[57]Bosselmann (2015), p. 195.

[58]The Whanganui River is New Zealand's longest navigable river and of special spiritual significance to New Zealand's indigenous people. This claim brought by local Iwi has been one of the longest and most fiercely contested in New Zealand history dating back to the late nineteenth

Crown, the other by Iwi to be known as 'the face of the river'.[59] The Te Awa Tupua (Whanganui River Claims) Settlement Bill was introduced into parliament on 2 May 2016, and is expected to be passed later in 2016. A list of the rivers intrinsic values is also recorded in the bill including that (a) the river is a source of 'spiritual and physical sustenance', (b) the river is 'an indivisible entity', and (c) the local Iwi have 'an inalienable connection to the river'.[60]

Christopher Stone has observed that human societies seem to evolve through certain stages of development and the law itself participates in this process.[61]

> Throughout legal history, each successive extension of rights to some new entity has been, theretofore, a bit unthinkable. We are inclined to suppose the rightlessness of right less 'things' to be a decree of Nature, not a legal convention acting in support of some status quo...The fact is, that each time there is a movement to confer rights onto some new 'entity', the proposal is bound to sound odd or frightening or laughable. This is partly because until the right less thing receives its rights, we cannot see it as anything but a *thing* for the use of 'us' - those who are holding rights at the time.

Perhaps the way forward is *more* sovereignty through environmental trusteeship and at the same time, *less* sovereignty in the sense of states becoming prepared to be bound by their trusteeship obligations through a World Environment Organisation. Its functions carefully constituted, could level the playing field between the politically strong and weak states and between the world market place and the environment, opening up a space for the creation of an emerging sustainable world system. Esty and Ivanova note that, "an extraordinary mix of political idealism and pragmatism is required to strengthen the international environmental regime".[62] The ideas expressed in this chapter of trusteeship and cooperation will be viewed by some as utopian, yet given the reality and seriousness of the issues, which threaten security and existence, we must demand that the political paradigm of environmental governance be transformed.[63]

Reimagining state sovereignty to accept limits to jurisdiction and to act as trustees for humanity under international standards of strong environmental institutions is the start of reclaiming and sustaining the commons we are so rapidly losing.[64]

---

century. The Waitangi Tribunal found in a report in 1999 that the Crown had taken control and possession of the river in breach of the Waitangi treaty and the crown's obligations.

[59]Jones (2015).

[60]This promising example of stewardship of a particular part of nature at the local and national level was hard fought for and the result of unwavering pressure on the New Zealand government by the local Iwi combined with early unlawful colonial activity recognized by the New Zealand courts along with a blending of Maori and western cultural ideas and concepts during the negotiation process. See Barraclough (2013); Harris (2015); Morris (2009); Hardcastle (2014), p. 6.

[61]Stone (1996), pp. 3–5.

[62]Esty and Ivanoa (2001), p. 18.

[63]Bosselmann (2015), p. 263.

[64]Ibid. at 270.

8 Conclusion

Protecting and restoring the integrity of ecological and human systems must guide the way we design our legal and institutional framework. This foremost requires an understanding why existing arrangements are ineffective and insufficient, in particular an appreciation that the mismatch between global systems and the different interests of diverse states is systemic and not just politically motivated.

At its core lies a change to the Westphalian concept of state sovereignty. In a globalized world, the state can protect its citizens only by incorporating global issues into national policies. With respect to economic, corporate and financial forces, this requires 'more sovereignty' in order to control them. With respect to the global commons, this requires 'less sovereignty' in order to adopt trusteeship functions. Such seemingly paradox approach, in reality, reveals 'smart sovereignty'[65] as it redefines territorial sovereignty simultaneously as a right to use and as a responsibility to protect. In this way, the democratic state may have a chance to regain some the legitimacy that has been lost in recent times.

References

Archibugi D (2002) Demos and cosmopolis. New Left Rev 13:27

Barraclough T (2013) How far can the Te Awa Tupua (Whanganui River) proposal be said to reflect the rights of nature in New Zealand? (LLB (hons) University of Otago

Biermann F (2011) Reforming global environmental governance: the case for a United Nation's Environmental Organization (UNEO). In: Stakeholder forum, earth system governance project, and VU University Amsterdam, The Netherlands, p 10

Bosselmann K (2015) Earth governance: trusteeship of the global commons. Edward Elgar, Cheltenham, p 9

Bosselmann K (2016) Governing the commons: can states be trustees? In: Westra L, Gray J, D'Aloia A (eds) The common good and ecological integrity: human rights and the support of life. Routledge, Milton Park, Abingdon, pp 270 and 271

Coghill K, Smith T, Samford C (eds) (2014) Fiduciary duty and the atmospheric trust. Ashgate, Farnham, Surrey England

Criddle EJ, Fox-Decent E (2009) A fiduciary theory of Ius Cogens. Yale J Int 34:350

Deneen PJ (2005) Democratic faith. Princeton University Press, Princeton, p xvi

Engel JR (2010) The earth charter as a new covenant for democracy. In: Bosselmann K, Engel JR (eds) The earth charter: a framework for global governance. KIT Publishing, Leiden, pp 29–40

Esty DC, Ivanoa MH (2001) Making International Environmental Efforts Work: The case for a global environmental organization. Presentation at the open meeting of the Global Environmental Change Research Community, Rio de Janeiro, October 6–8 2001, p 4

Hardcastle L (2014) Turbulent times: speculations about how the Whanganui River's position as a legal entity will be implemented and how it may erode the New Zealand landscape. Sir Edward Taihakurei Durie essay competition, Maori Law Rev, p 6

Hardin G (1968) The tragedy of the commons. Science 162(3859):1243–1248

[65]Kaul (2013), pp. 34–58.

Harris RC (2015) The changing face of co-governance in New Zealand. LLM, University of Canterbury
International Council for Science (ICSU) (2010) Earth system science for global sustainability: the grand challenges. ICSU, Paris, p 5
Jóhannsdóttir A (2009) The significance of the default, a study in environmental law methodology with emphasis on ecological sustainability and international biodiversity law. Doctor of Law jur dr, Uppsala University
Jones KC (2015) Legal personality for the Whanganui River. LLM Research Paper, University of Auckland
Kaul I (2013) Meeting global challenges: assessing governance readiness. In: Hertie School of Governance (ed) Governance Report 2013. Oxford University Press, Oxford, pp 34–58
Kim RH (2013) Unraveling the maze of multilateral environmental agreements: a macroscopic analysis of international environmental law and governance in the Anthropocene. PhD Australian National University, p 89
Meadows DH (1999) Leverage points: places to intervene in a system. Sustainability Institute
Morris JDK (2009) Affording New Zealand rivers legal personality: a new vehicle for achieving Maori aspirations in co-management? LLM, University of Otago
Redgwell C (1999) Intergenerational trusts and environmental protection. Juris
Redgwell C (2005) Reforming the United Nations trusteeship council. In: Chambers WB, Green JF (eds) Reforming international environmental governance: from institutional limits to innovative reforms. United Nations University Press, Tokyo, p 179
Sand PH (2013) The rise of public trusteeship in international environmental law. Third International Haub Prize Symposium, Murnau, 2013
Sax JL (1969) The public trust doctrine in natural resource law: effective judicial intervention. Mich L Rev 68:471
Stone CD (1993) Defending the global commons. In: Sands P (ed) Greening international law. Earthscan, London, p 34
Stone CD (1996) Should trees have standing? And other essays on law, morals and the environment, 25th edn. Oceana, New York, pp 3–5
Taylor P (1998) An ecological approach to international law. Routledge, London
Taylor P (2016) The common heritage: constructive utopianism In: Magalhães P, Steffen W, Bosselmann K, Aragão A, Soromenho-Marques V (eds) The SOS treaty: a new approach to managing our use of the earth system. Cambridge Scholars Publishing, Cambridge, pp 104–130
Weiss EB (2012) The coming water crisis: a common concern of Humankind. Transnational Environ Law 1(1):153

Part II
Human Security, Food and Water Issues

Pathways to Improved Water Law and Governance: Public Interest Litigation and Protest

Janice Gray

1 Introduction

Whether water should be characterised as a commons is increasingly the subject of discussion amongst members of civil society, policy makers, legislators and environmentalists, for example (Gray, forthcoming 2017). The discussion generally observes that: all biota is water dependent; water resources exist globally and are often interlinked; actions and conduct in one jurisdiction may have impacts on the quality and quantity of water in other jurisdictions; and water is a finite resource. Yet, with the exception of the high seas; some inter-national water basin agreements, and the (limited) impact of certain international law instruments, terrestrial water has largely been governed on a domestic basis with individual states being responsible for the water within their jurisdiction.[1] Such domestic approaches have rarely satisfactorily addressed the interdependence on water of global communities. Commonly national and other rivalries have driven domestic law and governance in the water domain.

Yet, awareness of issues such as the precious nature of water, its finite limits and the interconnectedness of different water resources highlight the importance of domestic jurisdictions being able to design and implement effective water law and governance. In other words 'getting water law and governance right' in domestic

[1]In international law the high seas, the atmosphere, Antarctica and outer space are regarded as global commons. The high seas have been subject to the *mare liberum* (freedom of the sea) doctrine. Some bi-national and multi-national agreements have also been entered into in relation to water management for example, in relation to the Mekong River Basin and some international law instruments may, in limited circumstances, impact on terrestrial waters (see *Tasmanian Dams* case). Terrestrial waters include surface water (such as rivers, creeks, lakes and lagoons billabongs) and groundwater (including water in aquifers and aquitards).

J. Gray (✉)
Faculty of Law, University of New South Wales, Sydney, New South Wales, Australia
e-mail: j.gray@unsw.edu.au

L. Westra et al. (eds.), *The Role of Integrity in the Governance of the Commons*,
DOI 10.1007/978-3-319-54392-5_5

jurisdictions becomes highly significant because that water law and governance will have impacts well beyond the domestic jurisdictions in which it is made. Given the universal dependence on water it is also not difficult to frame the preservation of water quality and the equitable allocation of water resources as being elements of the common good. Water's protection, therefore, has a strong ethical element and not to protect water is ultimately to contribute to the demise of both human and other security.

The capacity of states to effectively govern water is arguably not assisted by the absence in much domestic law of international law principles such as, or akin to, the 'common heritage of humankind'. That principle provides over-arching guidance; guidance which is not always as clearly enunciated by other principles in domestic legislation and policy. How then is effective water governance to be achieved? One method relies on continual governance refinements which emerge as necessary in the face of review and interrogation of existing law and governance, particularly as it relates to ecological sustainability. Yet, such improvements are only likely to occur when an awareness of problems (real, potential, individual and cumulative) exists and/or when opportunities are made available for courts to test the integrity of procedural and substantive issues relating to water law. Bringing concerns about water governance to public, governmental and judicial attention is, therefore, highly important.

This chapter takes one jurisdiction, Australia (and at times, more specifically New South Wales (NSW)) as the focus of discussion about (a) the effectiveness of water law and governance and (b) paths to improvement of that law and governance. It argues that public interest litigation may be employed to bring water law concerns to the court. However, there are several factors which may impede the commencement and/or passage of public interest litigation. It also argues that protest is an important means by which water concerns may be brought to public and political attention.

In its consideration of the ways in which deficiencies in water law and governance are brought to the fore it also explores two controversial legal reforms: one which would impact on the capacity of environmental organisations to commence litigation and the other which may deter individuals and groups from highlighting water concerns through the use of protest. Both reforms may, if implemented, militate against environmental (including water) protection and arguably act as disincentives for public interest litigation to proceed; litigation which often plays a vital role in helping protect water and the environment more generally.

2 Background

2.1 Australian Water Law and Governance

Water law is a key aspect of environmental and natural resources governance and Australia (the world's driest continent except for Antarctica) has been at the cutting edge of water law, policy and management for some years. Around the turn of the twentieth century State and Territory[2] legislatures across Australia (a federated system) undertook a series of comprehensive reforms which introduced new, legal, regulatory and institutional frameworks. Those frameworks have at their core comprehensive water planning and water licensing systems which support water trading. A cap and trade scheme is regarded as a key mechanism for sharing water resources and consequently legislation in each State (and Territory) sets out the range of tradable instruments and the manner in which trading may occur.

The following offers a brief overview of the way domestic water law operates so that the later discussion about water law's weaknesses may be better contextualised and understood.

In order to access water from surface or groundwater sources a water entitlement or water allocation is needed. A water entitlement is a perpetual share in a variable consumptive pool. The volume in the pool of water varies according to precipitation, capture, run-off and evaporation, for example. Hence in years of heavy rainfall there will be more water for the relevant Minister to make available to holders of entitlements but the share (or percentage) of entitlement will not vary, just the volume of water in the consumptive pool.

An entitlement holder may sell his/her entitlement in accordance with the provisions of the relevant State's water legislation and its water planning rules. Once sold, all interest in the entitlement is lost. However, if the entitlement holder does not wish to take such a drastic step and transfer his/her entitlement permanently, he/she may merely sell the allocation. The allocation is the actual volume of water that the entitlement makes available in any given water year. Hence the sale of an allocation is a temporary form of transfer while the sale of an entitlement is permanent. It is through transfers both permanent and temporary that sales are effectuated.

Since the execution, in 2004, of the National Water Initiative (NWI), which is the national blueprint for water reform, some aspects of water law and governance have additionally been imbued with a federal overlay (Gray 2012).[3] The federal

---

[2]Australia has six states (New South Wales, Victoria, Queensland, Tasmania, South Australia and Western Australia) and two territories (the Australian Capital Territory and the Northern Territory).

[3]The NWI was signed by all Australian states in 2004 (through the Council of Australian Governments) with the exception of Tasmania and Western Australia who became parties to the Agreement in 2005 and 2006 respectively.

overlay largely applies to water in the Murray Darling Basin (MDB) which is the food bowl for much of Australia.

The MDB reaches into the eastern states of Queensland, NSW, Victoria and South Australia and trade within and between the Southern MDB States is a key plank in the (NWI) water reforms mentioned above. That trade is, in part, governed by Commonwealth legislation and related instruments.[4] Hence, where relevant, the Water Act (Cth) 2007), the MDB Plan and the Australian Consumer and Competition Council's trading and market rules, for example, may also apply. Water law will, therefore, at times, necessarily need to operate trans-jurisdictionally which has proved a challenge for legislators, policy makers and others alike (Gray et al. 2016).

One of the propositions behind post NWI water law and governance is that trading should be supported and that moving significant volumes of water from one place to another is either a positive, or at least an acceptable, practice. The section below on weaknesses in water law and governance raises some concerns about that proposition although it is conceded that because the Australian hydrological cycle has 'the highest global variability for any inhabited continent, of rainfall distribution and run-off, in both spatial and temporal terms' effective water management is very difficult to achieve no matter the method employed (Gardner et al. 2009, p. 5). The pattern of human intervention which creates an overlay on these natural rainfall patterns magnifies the difficulties.

Meanwhile another significant NWI water law reform was the introduction of Sustainable Diversion Limits (SDLs). SDLs are designed to limit the volume of 'water take' to sustainable levels. The concept of Environmental Water and the institutions of State and Commonwealth Environmental Water Holders were also introduced as part of the NWI reforms. These mechanisms involve water being returned to the environment and views on their effectiveness are also raised below.

Two further issues, among others, with which Australian water law and governance has been grappling are: (a) water quality associated with unconventional gas mining projects (Gray in Gray et al. 2016) and; (b) diffuse source pollution particularly as it affects the Great Barrier Reef, for example (Waschka and Gardner 2016 in Gray et al. 2016).

3 Weaknesses in and Concerns About Water Law and Governance

Although the post NWI reforms have arguably assisted in alleviating the water over-allocation problem which beset (and continues to beset) Australia, the NWI water reforms have also revealed some weaknesses: weaknesses which may benefit from being brought to public and other attention through protest or resistance as well as from being explored more fully in the courts. However, as the purpose of

[4]Williams and Kildea (2010), pp. 595–616.

this chapter is not to analyse in any depth the weaknesses themselves but rather to consider the *ways* in which such weaknesses may be brought to public, judicial and other attention so as suitable reforms may ultimately ensue, this section is fairly cursory. It simply outlines the types of concerns associated with domestic water law and governance in Australia.

One weakness in water governance relates to the ways in which water trading and the market impact on the communities from which water is transferred out. The movement of water may, for example, lead to heightened water impoverishment[5] and a decline in the sustainability of some rural communities (Masterman-Smith et al. forthcoming 2016; Gray 2006, 2012). The impoverishment associated with water movement may also be exacerbated by climate change.

Climate change may encourage water movement because climate change results in rising temperatures and those temperatures contribute to water scarcity. As a result some communities suffering water scarcity may need to 'buy in' supplementary water in the future. This means that water is lost from another community. The depletion of water resources from some communities for delivery into others, via water movement through trade, may have implications for the 'supplying' region's capacity to respond to hazards, such as fight bushfires (a common Australian summer hazard). Scientists and others have also argued that by reducing water flow through trade (out of a region) and the consequent water movement, some areas may experience a rise in water particulate pollution.

Further, water scarcity and the movement of water out of certain locations have also been associated with increased river salinity in some areas (Gardner et al. 2009, ch 25). Some commentators argue that these and other impacts are likely to affect Indigenous communities more than others, leading to even greater water injustice (McLean 2007).

Whether the law governing the above areas is adequate to protect water resources and biota more generally deserves attention as does the question of the nature of the rights in access to water. For the security of water transactions it would also seem necessary to establish clearly whether water rights constitute property. (The remedies for a breach of property rights are usually more extensive than for breaches of many other kinds of rights.) However, space limitations prevent further discussion of those issues here.

As suggested above, the tension between environmental water rights and private water rights is also emerging as another major concern. Irrigators and agriculturalists argue that because water is being returned to the environment there is less water available in the variable consumptive pool (to be shared between water entitlement holders). The reduced volume of available water has the effect of pushing up the market price of water and if that price continues to rise, irrigators and agriculturalists argue that they may no longer be able to afford water to grow the crops necessary to promote food security.

---

[5]Water impoverishment here refers to the decline in water availability in some areas that leads to further socio-economic inequality.

Associated with the environmental water issue is the issue of SDLs. Much angst still exists around the level at which SDLs in the MDB is set but the concept of water trading more generally has also come in for criticism amongst many irrigators who argue their responsibility for food security ought permit them to receive non-market related water entitlements (Hanjra and Qureshi 2010; Sullivan 2016). Without such protection from market fluctuations prime farm land is now 'dead' according to some (Sullivan 2016).

Yet another water governance concern relates to water quality associated with the burgeoning number of unconventional gas mining projects. Unconventional gas (including coal bed methane, shale gas and tight gas) mining and the associated technique of hydraulic fracturing (fracking) have been the subject of criticism because those activities, it is argued, may contribute to degraded water quality through the introduction of chemical fracking fluids, the release of naturally occurring toxic chemicals formerly securely secreted underground, cross-aquifer contamination, a failure of well integrity and the use of evaporation ponds, for example (Pramod 2011; Bately and Kookana 2012; Hoffman 2016). Consequent seismic activity associated with fracking is also of concern as it may indirectly compromise water quality by, for example, opening up fissures and cracks between contaminated and non-contaminated groundwater reserves.

In the USA there is an established connection between fracking and environmental (and other) damage (Hoffman 2016).[6] Similar concerns about the connection between fracking and environmental damage also exist in Australia and are detailed elsewhere (Gray 2014).

4 Recognising Weaknesses and Improving Water Governance: Public Interest Litigation and Protest as Tools for Enhancement

In this section we proceed on the basis that public interest litigation could potentially not only help improve water governance by clarifying, challenging or causing change to existing water law but also by bringing water law concerns to judicial and public attention in the first place. Yet, there are several factors which, if not properly addressed, may act as disincentives to the pursuit of public interest litigation. A selection of those factors is discussed below.

[6]The Environmental Protection Agency (EPA) and the United States Geological Survey, for example, confirmed that fracking had contaminated groundwater. As a result, the EPA, under an emergency administrative order, required three oil production companies with operations on the Fort Peck Reservation to reimburse the city of Poplar for expenditure that city incurred in relation to water infrastructure as the result of drilling contamination. Although the oil companies appealed the EPA order, a Federal judge required them to rectify their violations (Hoffman 2016).

We also consider how protest and resistance may be used as tools to bring weaknesses in water law and governance to public and political attention. We go on to observe, that opportunities to rely on these tools may be closed off by legal reforms in NSW which have the potential to disempower and deter those seeking to highlight water-related (and other) concerns through protest. A more detailed discussion of the reforms affecting protest is undertaken in the section on 'Anti-protest laws in NSW'.

4.1 Public Interest Litigation

To date there has been some water litigation in Australia (particularly NSW) but most water litigation has been private litigation rather than public interest litigation. There has, for example, been litigation related to: the construction of structures for diversion, retention and the conveyance of water; the impact of the hydrological scheme on development; the use of water by development (Preston 2009) and the 'acquisition of property' in the form of bore licences (*ICM Agriculture Pty Ltd. v Commonwealth*).

The push in favour of public interest litigation is, however, gaining momentum but as Chief Justice Preston, of the NSW Land and Environment Court, observed extra-curially, 'the growth in the numbers of [public interest] actions has not been matched by an increase in success' (Preston 2013).[7] This begs the question, why is it so difficult to use public interest litigation effectively as a tool of water protection? What are the disincentives or barriers to water-based public interest litigation? The answer to that question is that the disincentives which militate more generally against environmental public interest litigation apply equally to water-based litigation (a sub-set of environmental litigation). Using the Preston framework, we consider those disincentives below (Preston 2013).

4.1.1 Adequate Water Laws

It is necessary to establish a 'foundational' water law in which to ground public interest litigation. As Preston observes, if there is no right of action, then there is clearly no possibility of litigation (Preston 2013). Without an effective foundational water law, we do not even get to first base in our quest for improved water justice.

It is also necessary that the relevant legislation: contains provisions which are able to restrain and remedy non-compliance; allows for compensation to be

[7]This section on Public Interest Litigation draws heavily on Justice Brian Preston's extra-curial writings. It relies on his framework for discussion of public interest litigation and condenses and summarises many of his ideas. The author is indebted to His Honour. Any errors in this section lie with the author rather than His Honour.

obtained for loss or damage caused by a breach of duties; has the capacity to permit review of the merits of administrative decisions on a re-hearing (a merits' review) and; possibly has the availability of criminal actions so that water wrongdoers may be prosecuted or punished for their offences against water-related laws (Preston 2013, p. 1). If public interest litigation is to succeed in the water sphere, the relevant law underpinning it, therefore, needs to cover the breadth of issues which may arise. Further the law must be both effective and enforceable. Hence, if the law precludes the standing of some potentially relevant parties or denies suitable remedies which would redress injustice, it may be said to fail the test of adequate foundational law. (Standing is discussed in more detail below.)

4.1.2 Justiciability

For public interest water litigation to play an effective role in improving both water quality and quantity the issue of concern must be justiciable. This means that the issue must be one capable of being considered legally and determined by the courts. Hence if the issue cannot be brought under the ambit of existing legislation or common law then that issue is not justiciable and no matter how unjust an activity or action may be, the Court cannot hear the matter.

In the water sphere, questions of justiciability may arise in many circumstances but particularly in relation to international (water or water-related) obligations under international law treaties, covenants and other instruments. However, a breach of a country's obligations under international law will not alone cause that issue to be justiciable in a domestic court at the suit of a private citizen (see *Tasmanian Wilderness Society Inc. v Fraser*). But if obligations under international law are transposed into domestic law, then it is possible that the relevant issue is justiciable under that domestic law if a remedy is also available (see Preston 2013, p. 5).

It is, therefore, the case that in the water context as indeed in other contexts, harm, unconscionability or injustice, for example, will not alone create the conditions in which a suit can be brought. This issue is sometimes misunderstood by members of the public who observe injustice or outcomes that affect the environment detrimentally and are keen to see the situation remedied.

4.1.3 Willing and Able Litigants

One of the most difficult barriers to overcome in bringing public interest litigation is finding willing and able litigants. Litigation is not only costly but also requires a strong commitment to the cause of that litigation. Preston astutely notes that:

> [w]illingness is a product of not only enthusiasm and zeal for the environmental cause but also a cultural attitude. The cultural tradition of the country needs to support, and not inhibit or punish, citizen access to justice through bringing legal suits in the courts. Public interest litigation is often a form of protest, challenging powerful interests in the government and the private sector (Preston 2013; p 7).

Hence litigant resilience is required along with a cultural context that is conducive to such litigation. A certain level of tolerance for difference is, therefore, a pre-condition to the pursuit of effective public interest litigation. Public interest litigation will clearly be more robust in jurisdictions where avenues to pursue it are left open. In those jurisdictions public interest litigation may be characterised as a tool which helps serve the common good. It may promote a healthy, inclusive and participatory democracy and it may involve the promotion of free speech and access to justice; attributes which are compatible with serving the common good. Yet, some Australian and NSW law reforms will arguably have the effect of shutting down some public interest (water) litigation opportunities. They are discussed below under the heading of 'Reforms and disincentives to public interest (water) litigation'.

Another pre-condition to locating willing and able litigants is ensuring that potential litigants are well-versed in the relevant subject matter and that they understand (at least to some extent) the substantive and procedural law that governs the suit. Whilst such people may not need to be experts, a good working knowledge of the way law does or may, operate in relation to the relevant subject matter is a great advantage.

4.1.4 Appropriately Skilled Lawyers

Water litigation is more likely to succeed if litigants are represented by appropriately trained and skilled lawyers. While litigants may be passionate about their cause, the finely grained knowledge which a skilled lawyer is able to bring to the case, will enhance that case's chances of success (see Sarat and Scheingold 1998). As environmentally-based public interest litigation may take a long time to work its way through the judicial system, continuity of the legal team is also important. Ludwig Kramer has noted the legal, institutional and factual memory of the legal team is likely to prove significant in building a cohesive and convincing evidence-based case (Kramer 2014).

4.1.5 Cost of Litigation: Who Can Fund Public Interest Litigation?

All litigation is expensive and accordingly it is often difficult for those wishing to pursue public interest litigation to find suitable ways to fund it. One method to which Preston refers is self-funding (Preston 2013). As the plaintiff in environmental litigation is commonly a citizen group it is likely that such a group will often seek to rely on philanthropic donations and fund-raising activities to support its 'self-funded' litigation. Such an approach has not proved particularly successful in Australia which (unlike the US) does not enjoy a strong philanthropic tradition more generally. Further, the adversarialism in which Australian litigation is rooted has arguably also been a deterrent to philanthropic support (Pring and Pring 2009 as cited in Preston 2013).

Other funding models include crowd-funding. Although this method may sound attractive, it has not yielded a great deal of success in some other Australian jurisdictions. For example, in the Northern Territory (NT), the Environmental Defender's Office [EDO] did not manage to raise enough to fund the salary of one lawyer (Australian Parliamentary Business Committee 2015). (It only raised $1,0,000.00). As the public becomes more acculturated to this mode of funding, it is possible that the method's success may improve but that has not yet happened.

Reliance on government legal aid to fund public interest litigation is another option but in NSW (and indeed many other Australian jurisdictions) it is becoming increasingly challenging to obtain such aid. Not only must the case have merit (that is, reasonable prospects of success) but the plaintiff also needs to satisfy a strict means test.

Water litigation funding is also possible on a pro-bono basis. However, given that many of the solicitors and barristers who could potentially undertake such work may also represent defendants, the pro-bono option has not proved a very successful funding model in Australia for environmental (including water) litigation. The likelihood of pro-bono funding taking a stronger hold in water sector litigation is perhaps also affected by a downturn in the work of private, commercial law firms in Australia (see Papadakis 2016).[8] Presumably such a decline will impact on the extent of pro-bono work more generally.

EDOs theoretically offer another avenue for funding public interest (water) litigation but funding cuts to them and the tenuous nature of their existence have not instilled confidence that such institutions will be able to take up important public interest (and perhaps long-running) water litigation. In December, 2013, for example, the then Federal (Liberal/National Party) government[9] cut 10 million in funding over 4 years, for nine States' and Territories' EDOs which specialise in public interest environmental and planning law. The cuts came without warning (EDO 2013). Not only did the Liberal/National Party government tend to imply that much of the litigation undertaken by EDOs was vexatious, frivolous and/or unnecessary (engendering the view that such litigation should not be supported) but it also came to light that the government was subject to pressure from key business-related forces to cut the funding. A Freedom of Information application revealed that the Minerals Council of Australia (a key mining lobby) had written to the then Attorney General (Senator Brandis) asking him for funding cuts to EDOs (Robinson 2015).

[8]In many cases this has led to partners being forced out of their existing positions. The trend towards long hours in a competitive environment may also impact on lawyers' willingness to provide pro bono services. On the competitive nature of legal practice see Thornton (2016). Further some law firms are predicting that they will be able to use technology to such an extent that they will need to employ far fewer junior solicitors, an outcome which would have the effect of decreasing the number of lawyers available to undertake pro bono work.

[9]That is, the government led by Prime Minister, Tony Abbott.

The same coalition government also threatened to make donations to EDOs no longer tax deductible. In giving evidence to the Australian Senate Hearing on the Federal government's push to end the tax deductible status of donations to (environmental) legal centres—environmental legal centres' work often involves challenging the interests of big business—Rachel Walmsley, an EDO Policy and Reform Director observed that 'attacks on the environment' would force the Northern Territory EDO to close within weeks and the South Australian and Australian Capital Territory EDOs would be forced to follow within six months (Robinson 2015). Clearly institutions such as EDOs cannot be confidently relied upon as being able to undertake public interest (water) litigation when their very existence is so tenuous.

The governmental conduct described above goes to the heart of Justice Preston's concerns quoted in the discussion under the heading '[w]illing and able litigants'. The conduct described reveals a culture not open to reflection and criticism. It reveals a culture keen to shut-down dissent and disagreement.

4.1.6 Locus Standi

Public interest litigation is dependent on the plaintiff having locus standi (standing). Standing depends on the identity of the person, the type of subject matter of proceedings, and the relationship between the person and those proceedings. In Australia 'the common law test for standing to bring judicial review or civil law proceedings is that the person can sue, without joining the Attorney General, in two cases: first where the interference with the public rights is such that some private right of the person is at the same time interfered with and, secondly, where no private rights is interfered with, but the person has a *special interest* [emphasis added] in subject matter of the action' (Preston 2013, p. 14; *Boyce v Paddington Borough Council* reformulated in *Australian Conservation Foundation v Commonwealth* p. 527).

The special interest needs to be more than intellectual or emotional but it is not required to be pecuniary or proprietary (*Australian Conservation Foundation v Commonwealth* pp. 530–531; 539, 547–548). The special interest requirement may, therefore, act as a gate-keeper as to who can bring public interest (water) litigation. Active use of the land which is the subject of litigation (*Fraser Island Defenders Organisation* v *Hervey Bay Town Council*); an adverse impact on the amenity of the plaintiff's land (eg *Onus v Alcoa (Aust) Ltd*); protection of statutory participation rights (eg *Sinclair v Maryborough Mining Warden*; *Australian Conservation Foundation v Forestry Commission of Tasmania*; or, cases where a government decision or the conduct challenged relates to the objects and activities of the plaintiff organisation (*Australian Conservation Foundation v Minister for Resources*) may amount to a special interest (Preston 2013, p. 14).

At the State level in NSW, environmental legislation generally contains open standing provisions meaning that any person (natural or legal) has standing to bring proceedings. Such standing also allows any person (natural or legal) to restrain a breach of legislation. In both cases it is not necessary for a right of that person to have been infringed (or for that person to demonstrate that it may be infringed) by or as a consequence of the breach (*Environmental Planning and Assessment Act 1979 (NSW)* s 123).

Under Commonwealth legislation 'green' standing is given, for the purpose of seeking injunctions relating to conduct or proposed conduct in contravention of the *Environment Protection and Biodiversity Conservation Act* (EPBCA), to individuals and organisations whose interests have been or would be affected by the conduct in question, or to those individuals and organisations who have engaged in activities in the previous two years, for the protection or conservation of, or research into the environment, or in the case of an organisation whose objects or purposes include the protection or conservation of, or research into, the environment (EPBC s 475(6)). Such a standing provision is well regarded amongst environmentalists and many others because it does not deliberately place barriers in the path of public interest environmental litigation.

4.1.7 Other Aspects in Bringing Public Interest Litigation

As with all litigation the parties litigating water issues are dependent on proof established by evidence. In public interest litigation the plaintiff bears the onus of proof but it may be difficult for the environmental (including, water) plaintiff to collect sufficient high quality evidence to prove his/her case for a number of reasons. Water experts may, for example, resist sharing documents and other material with plaintiffs if those experts are commonly employed by governments as consultants. They may not wish to jeopardise the prospects of future government work by supplying evidence to the 'other side'.

One tool that public interest litigants may employ relatively easily is a freedom of information application. Applications made under legislation such as the *Government Information (Public Access) Act* 2009 (NSW) GI(PA)A cause government agencies and institutions to divulge information in certain conditions. In NSW, the Land and Environment Court also has the capacity to direct a public authority to make available to the plaintiffs any document that records matters relevant to the decision. The GI(PA)A along with the Court's power to direct, represent a positive step in the promotion of transparency and are, therefore, helpful tools of public interest litigation. They may be seen as promoting the common good and indirectly aiding the protection of resources such as water; a resource that needs to be stewarded well for the benefit of all biota (and simply for its own benefit).

Public interest (water) litigation will also be affected by the interlocutory practices and procedures in the relevant jurisdiction and by issues such as delays in the hearing of cases. For all litigation including public interest litigation to serve society well there also needs to be an adherence to the separation of powers doctrine which maintains separation between the judiciary, legislature and executive. The judiciary, in particular, needs to be able to act independently, free from the influence of the legislature and others who may seek to pressure the legislature to influence the judiciary and court process.

In Australia there has been an increasing number of attacks on the judiciary by outspoken and rabid radio commentators, for example, who have developed a reputation for deriding judicial decisions with which they do not agree. The growing trend for talk-back radio personalities, colloquially known as 'media jocks' or 'shock jocks', to engage in the (often ill-informed) denigration of judicial decisions is a concerning feature of Australian society. While freedom of speech is generally a positive feature of a robust democracy so is the preservation of democracy's key institutions including the judiciary. Preservation of the judiciary is dependent on respect. When attacks deride judicial decisions which are legitimately made by application of current law, a foundational institution of Australian democracy is threatened. Some of the attacks on judicial decisions have related to water, particularly water in the mining (including coal seam gas mining) context.

5 Reforms and Disincentives to Public Interest (Water) Litigation

In this section we discuss two reforms (one mooted and one actual) both of which may deleteriously impact on the protection of water because they have the potential to deter public interest water litigation.

One is a proposed amendment to the *EPBC* and the other is a series of amendments originally contained in the Inclosed Lands: Crimes and Enforcement Legislation Amendment Interference Bill (NSW) 2016.

5.1 EPBCA *Reform*

Section 487 of the *EPBCA* provides extended standing for judicial review. In its current form and as noted above, it allows actions to be brought 'by a person aggrieved by the decision' and it extends that definition to include individuals and organisations which have engaged in 'a series of activities related to the protection or conservation of, or research into, the environment' within the two years preceding the decision (see EDO 2015). The provision was originally introduced so environmental and conservation groups could sue on behalf of affected communities even when the environmental and conservation groups were not part of those communities. The rationale for its introduction was to support environmental litigation without going as far as introducing open-access standing. The section has been characterised as having introduced 'representative standing' in order to promote rigorous environmental reviews and approvals.

Section 487 has been employed, on a number of occasions, in order to bring environmental actions. It was, for example, relied on by the Mackay Conservation Group to initiate legal action against the approval of the Adani Group's Carmichael

Coal mine. The conservation group was successful in its appeal against the Australian government's initial approval of the mine because the group was able to demonstrate that the Minister's decision was made without due regard to the relevant Act.

Partly in response to the conservation group's reliance on the provision (reliance which led to the cancellation of Adani's approval) the Commonwealth Attorney General, Senator George Brandis, led the charge to repeal s 487(2) of the *EBPCA*. Senator Brandis claimed that the existing provision 'allows radical green activists to engage in vigilante litigation to stop important economic projects' and 'provides a red carpet for radical activists wanting to use aggressive litigation tactics to disrupt and sabotage important projects' (such as coal mines and unconventional gas projects) (Brandis 2015; Clark 2015).

The Court's requirement that Adani follow due process appeared to incense the Federal government causing it to suggest this bold and controversial reform. The Brandis amendment, as it has become known, has been heavily criticised by members of civil society and is regarded by many as disproportionate to any grievance.[10]

Government reactions to the Adani case stand in stark contrast to reactions to the Metgasco case. In 2015, the unconventional gas mining company, Metgasco had its petroleum exploration licences cancelled (unconventional gas activities may, as we have noted above, affect water quality and quantity). Metgasco brought a suit (*Metgasco Ltd. v Minister for Resources*) in order to appeal the Ministerial decision cancelling its licences. It won the suit and the NSW government had to pay compensation for wrongfully cancelling the licences. However, the overturning of the Ministerial decision in the Metgasco case did not generate the same level of vocal disapproval from government politicians as did the overturning of Adani decision. Perhaps it can be concluded from this that conservation groups are seen as not having a right to challenge decisions but mining companies are seen as having such a right (Clark 2015).

Clearly one consequence of repealing s 487 would be to diminish severely the watchdog role of environmental groups and seriously curtail the ability of those groups to engage in public interest litigation for the protection of water and other resources. If the amendment to s 487 results in the denial of standing to conservation groups because the goals of those groups do not align well with government's plans for further natural resource exploitation and further economic expansion, this would constitute an attack on democratic participation. It would also introduce a barrier and a disincentive to the use of public interest litigation as a tool of water protection. Given water's domestic and global importance this would represent an unfortunate way to proceed.

---

[10]Redfern Legal Centre holds this view, for example. Note as a postscript, that fresh approval was eventually granted to Adani but not without the Minister first having to reconsider the matter and comply with due process.

Water- related cases such as *Western Downs Alliance v Minister for Environment* v *Santos Ltd.*[11] (presently being litigated by the NSW EDO) may never have reached the courts if the proposed amendments to the *EPBCA* provision were already in place. (*Western Downs Alliance* argues that the approval of 6000 coal seam gas wells was unlawful because the Minister did not properly assess the project's impact on surface water.)

In summary, there are several significant reasons to retain the extended standing provisions. They include: a general public interest in ensuring that decision-makers comply with legislative procedures; third party rights although not always used do provide additional support for public interest (water) litigation and 'promote improved decision-making, accountability and public confidence in law'; representative standing means that individual water (or land) holders do not bear all the responsibility themselves for protecting the environment; and broad, representative standing reflects a trend that is supported by recent reviews including a 2009 review of the *EPBCA* (see Australian Senate 2015). (Note the Hawke Review actually recommended expanding standing provisions.) While extended standing has played a key role in facilitating public interest litigation there is no evidence that it has opened the floodgates with courts (McGrath 2016). It should remain to support future water-based public interest litigation.

5.2 *Anti-Protest Laws in NSW*

Protest is an important means of bringing water (and other) concerns to public attention. It is commonly the precursor to public interest litigation, highlighting weaknesses in the operation of law and governance, and providing a climate in which public interest litigation planning may be fostered and effectuated. As with public interest litigation itself, protest is a means by which society interrogates and responds to actions and policies. Yet, in NSW, reforms which are likely to deter public protest in certain (water-related and other) circumstances have been passed by both houses of parliament and are in the process of being introduced.[12] The reforms which are contained in the Inclosed Lands: Crimes and Enforcement Legislation Amendment Interference Bill 2016

1. Amend... the Inclosed Lands Protection Act 1901 to create an aggravated form of the offence of unlawful entry on inclosed lands, increasing the maximum penalty from \$550 to \$5500.....

[11]This case tests the approval of more than 6000 new coal seam gas wells in the Surat Basin, Queensland, Australia. Coal seam gas mining both uses and produces very large volumes of water.

[12]At the time of writing, the Bill had passed both houses of parliament and aspects of it were awaiting assent.

2. Amend the Crimes Act 1900 in relation to the offences of intentionally or recklessly interfering with a mine to extend the meaning of 'mine' to 'mineral, gas or petroleum exploration sites'
3. Amend the Law Enforcement (Powers and Responsibilities) Act 2002 (LEPRA) to confer additional search and seizure powers (without warrant); and
4. Amend the LEPRA to remove limitations on the exercise of police powers to give certain directions in public places (Law Society of NSW 2016a).

At first blush these amendments do not appear to have water implications but further investigation reveals that some (particularly numbers 1, 2 and 4) may well impact on water. For example, the legislative amendments operate to prohibit the unlawful entry of protestors onto unconventional mining and other mining sites. As we have observed above both types of mining have implications for water health and water quantity. Indeed the motivation for some protest at these sites is likely to be a desire to protect water, with protestors who are concerned about what they see as a state usurpation of responsibility, casting themselves as the guardians of water quality (Gray 2016).

Further the very act of (water-related) protest, irrespective of whether it involves unlawful entry on inclosed lands, may lead to the police invoking their new and very extensive powers under the amendments, if there is concern about 'public safety'.

It has been said that the amendments were introduced in response to 'risky protesting activities ….caused by a small minority who have shown a clear disregard for the safety of themselves and others'. According to the Minister for Industry, Resources and Energy, 'by community standards there are some actions that clearly should not be supported. Threatening the safety of others and causing disruption to legal business activities are clearly in that camp' (Roberts 2016). Protestors have also been described as a 'radical minority' who 'abuse' the democratic right to protest. Meanwhile the new measures have additionally been portrayed as measures that respond to 'organised crime gangs' who threaten 'our way of life' (see discussion of McNamara and Quilter 2016). Whether a key driver for reform is actually a desire to curb organised crime gangs is perhaps contestable. Few would have difficulty with reigning in the influence of organised crime gangs but it is unlikely that water-related protest coming to the fore at unconventional mining sites is an arm of organised crime gangs.

The anti-protest amendments give a senior police officer, rather than a judge, the power to make 'public safety orders' which will prohibit the person against whom the order is made from attending a specified public event or entering specified premises for up to 72 hours (McNamara and Quilter 2016). Orders may be made by the police officer if he/she believes there is a 'serious risk to public safety or security'. Risk is assessed on the basis of whether the person's presence might lead to death or physical harm to a person, or serious property damage. The legislation requires the police take into account a range of matters which may include the person's criminal history and the reason for attendance at the relevant event. McNamara and Quilter observe that '[l]ots of cautious words and talk of "exemptions", such as whether the purpose of a person's presence is industrial action, don't change the fact that the proposed law gives the police very wide discretion to decide who should be allowed to circulate freely in public' (McNamara and Quilter 2016).

By contrast, the pre-amendment law in NSW permitted police to remove obstructions to traffic or people but prevented police from doing so in relation to industrial disputes, genuine protests or organised assemblies. By repealing s 200 of LEPRA police have been given a discretion as to whether they issue public safety orders. Repeal of the section, therefore, effectively removes important checks on police powers. Water (and other) protestors may, under the amendments, be ordered to 'move on' and/or face the prospect of a criminal record and a $550,000 fine leading commentators to conclude that '[t]he consequences [of the amendment] are serious' (McNamara and Quilter 2016).

In voicing its concern the NSW Bar Association noted that the amendments remove protections for political activities which have been 'properly regarded as an essential part of the social, political and cultural life of any civilised society' (Slezak 2016). Meanwhile the Law Society of NSW observed that it was uncertain as to 'why intending to interfere with a business should be an aggravating factor' (Law Society of NSW 2016b).

Under the amended law, a protestor who 'intentionally' or 'recklessly' interferes with a mine may be charged with aggravated unlawful entry on inclosed lands for which the maximum penalty is seven years' imprisonment. That penalty is greater than the NSW penalty for the offence of assault occasioning actual bodily harm. (Pursant to s 59(1) *Crimes Act* that offence carries a penalty of five years' imprisonment.) Hence a protestor seeking to raise awareness about prospective water degradation and chaining him/herself to a gate at an unconventional gas mining sites risks seven's year's imprisonment. Whether this position is commensurate with the seriousness of the offence is debatable.

Further, some may suggest that on one reading of Minister Roberts' Second Reading parliamentary speech in relation to the Bill, the word 'unsafe' may possibly be regarded as superfluous. This view is based on the belief that the real motivation for restricting the ability to protest is not so much a concern for safety as a concern to avoid interference to business activities; activities which are very much part of the NSW government's business plan for economic growth. On this view safety concerns become a mere (or almost, mere) justification for the restrictions.

Yet, protest is often what ignites public consciousness. It is one of the methods by which people become informed so that public interest litigation might later ensue. (Who knew about the ill-effects on water from fracking and unconventional gas mining before resistance and protest groups started forming?) Protest is, therefore, an important tool of democracy. It reflects a right to free speech and exemplifies an opportunity for engaged public participation which may ultimately contribute to the enhancement of (water) law and governance.

6 Conclusion

Water is a valuable resource that sustains life. In recognition of this and the need to share water across a global community of users there has been a push (in some quarters at least) to classify water as a global commons. To date, however, the only water to be classified as a global commons is marine water (the high seas). Terrestrial waters have not been so classified although some bi and multi-national agreements have been agreed with the aim of developing better trans-national and trans-jurisdictional approaches to water law and governance.

Terrestrial water remains largely governed on a domestic basis and accordingly water law is not necessarily guided by universal principles such as the over-arching, international law principle of 'the common heritage of humankind'. Without such guidance, the global community is dependent on individual nation states, such as Australia, or unions of nation states (such as the European Union) 'doing the right thing' by water law and governance and implementing law and policy which aim at long term sustainability, to ensure that enough water, of an adequate quality, is available for the global community now and into the future.

This chapter reviewed the framework for water law and governance, in one (cutting-edge) domestic jurisdiction, that of Australia (and more particularly, NSW), and it analysed pathways for improvement of that law and governance. In recognition of the role that public interest litigation may play in enhancing water governance, the chapter considered some aspects of public interest litigation in detail, bearing in mind that public interest litigation may bring water concerns to public and judicial attention; re-set the direction of water law; and ensure that current water law is upheld, for example. The chapter relied on Preston's framework (and many of his views) to conduct the discussion of factors which impact on how readily public interest litigation may be pursued. It, therefore, dealt with factors including the adequacy of laws, justiciability, the existence of willing and able litigants, the need for appropriately skilled lawyers, the cost of litigation and *locus standi*. It identified a range of issues which, if not adequately addressed, would act as disincentives to persons or organisations seeking to pursue public interest litigation.

The chapter took two significant legal reforms (one actual and one proposed) as the focus of specific discussion about the role of standing and the role of protest in effectuating legal change (and improvement) in the water sector. In particular, it examined proposed amendments to the *EPBCA* ; amendments which are designed to limit the representative standing presently enjoyed by environmental organisations; organisations (such as the NSW EDO) which bring water-related cases to court.

It also examined amendments arising out of the (now passed) Inclosed Land: Crimes and Enforcement Legislation Amendment Interference Bill 2016 pointing out how the amendments in that Bill could impact on the development of water law and governance by limiting the opportunities for, and the types of, protest permitted around water issues. It noted that (a) protest may be seen as a form of awareness-raising which often precedes the pursuit of public interest litigation and (b) although some of the reforms may initially appear to be unrelated to water they may, in fact, impact on water governance. This is so because some of the reforms apply to unconventional gas mining sites and one of the key concerns for unconventional gas mining protestors is how water quality and quantity are affected by unconventional gas mining activities.

What is clear is that terrestrial water throughout the world needs to be stewarded, maintained and used in such a way as to promote the preservation of adequate water quality and quantity for all who need it. This requires a vigilant application of robust law. Protest plays an important role in that pursuit in any civilised society as does ensuring that a wide range of persons or entities, who may wish to pursue a litigious course of action, for the benefit of the whole society, may do so. Measures to protect these (water) governance tools need to be maintained and strengthened rather than dismantled or limited.

References

Articles and Chapters

Australian Parliamentary Business Committees, Senate and Environment and Communications Committee, Australia's Environment Report, 'Environmental Defenders Offices', Chapter 3, 23 June 2015. Available http://www.aph.gov.au/Parliamentary_Business/Committees/Senate/Environment_and_Communications/Ausenviron/Report/c03. Accessed 10 Oct 2016

Australian Senate, Environment and communications legislation committee, environment protection and biodiversity protection (standing) bill provisions, November 2015

Bately G, Kookana R (2012) Environmental issues associated with coal seam gas recovery. Environ Chem 9(5):425–428

Brandis G (2015) 'Government Acts to Protect Jobs from Vigilante Litigants' (Media release, 18 August, 2015). Available at http://parlinfo.aph.gov.au/parlInfo/search/display/display.w3p;query=Id%3A%22media%2Fpressrel%2F4020386%22. Accessed 1 Nov 2016

Clark C (2015) The government vs the environment: lawfare in Australia. The Conversation. Available at https://theconversation.com/the-government-vs-the-environment-lawfare-in-australia-46205. Accessed 16 Oct 2016

EDO (NSW), Briefing Notes, EPBC Bilateral Agreement Implementation Bill, Before Senate Could Weaken National Environmental Oversight, Briefing Note, September, 2015. Available https://d3n8a8pro7vhmx.cloudfront.net/edonsw/pages/2282/attachments/original/1443054743/150924_EPBC_Bilat_Bill_-_briefing_note_FINAL.pdf?1443054743. Accessed 1 Nov 2016

EDO (NSW), EDO Offices Face Closure After Federal Funding Cuts. Media Release, 17 December 2013. Available. http://www.edonsw.org.au/edo_offices_face_closure_after_government_funding_cuts. 2 Oct 2013

Gardner A, Bartlett R, Gray J (2009) Water resources law. LexisNexis, Sydney

Gray J (2006) Legal approaches to the ownership, management and regulation of water from riparian rights to commodification. Transforming Cult E J 1(2): 64–96

Gray J (2012) The legal framework for water trading in the Murray darling basin: an overwhelming success? Environ Plan Law J 29(4):328–348

Gray J (2014) Frack Off! Law, policy, social resistance, coal seam gas mining and the earth charter. In: Westra L, Vilela M (eds) The earth charter, ecological integrity and social movements, Routledge/Earthscan, Oxon/New York
Gray J (2016) Trans-jurisdictional water law and governance in the context of unconventional gas mining: the Australian experience. In: Gray J, Holley C, Rayfuse R (eds) Trans-jurisdictional water law and governance. Routledge Earthscan, Abingdon
Gray J, Holley C, Rayfuse R (eds) (2016) Trans-jurisdictional water law and governance. Routledge Earthscan, Abingdon
Gray J (in press, 2017) Interview with Gray J on water as a global commons. Georgetown J Int Aff
Hanjra MA, Qureshi E (2010) Global water crisis and future food security in an era of climate change. Food Policy 35:365–377
Hoffman J (2016) Geology and Human Health, Topical Resources, 'Potential Health and Environmental Effects of Hydrofracking in the Williston Basin, Montana'. Available http://serc.carleton.edu/NAGTWorkshops/health/case_studies/hydrofracking_w.html. Accessed 20 Nov 2016
Kramer L (2014) Keynote address, IUCN Annual Conference, Tarragonna, Spain
Law Society of NSW, Submission to NSW Parliamentary Inquiry into the Inclosed Lands, Crimes and Law Enforcement Amendment (Interference) Bill 2016a. Available at http://www.lawsociety.com.au/cs/groups/public/documents/internetpolicysubmissions/1098132.pdf. Accessed 20 Nov 2016
Law Society of NSW, Submission to NSW Minister for Industry, Resources and Energy, The Hon Anthony Roberts MP, March 2016b. Available at http://www.lawsociety.com.au/cs/groups/public/documents/internetpolicysubmissions/1098132.pdf. Accessed 10 Oct 2016
McGrath C (2016) Myth drives Australian Government attack on standing and environmental 'lawfare'. EPLJ 33:3
McLean J (2007) Water injustices and potential remedies in indigenous rural contexts: a water justice analysis. Environmentalist 27:25
McNamara L, Quilter J (2016) Greater police powers and penalties threaten civil liberties in NSW for "public safety". The Conversation, 8 April 2016. Available at https://theconversation.com/greater-police-powers-and-penalties-threaten-civil-liberties-in-nsw-for-public-safety-57133. Accessed 1 Nov 2016
Masterman-Smith H, Rafferty J, Dunphy J, Gull Laird S (2016) The emerging field of rural environmental justice studies in Australia: Reflections from and environmental community engagement program. J Rural Stud 47: Part A: 271–277
Papadakis M (2016) Law Firms Shrink Partner Numbers as Clients Cut Back. Australian Financial Review Weekend, 23 June 2016. Available http://www.afr.com/business/legal/law-firms-shrink-partner-numbers-as-clients-cut-back-20160620-gpnvdm. 1 Oct 2016
Pramod T (2011) Coal bed methane production. In: Darling P (ed) SME mining engineering handbook. Society for Mining, Metallurgy and Exploration, Englewood
Preston B (2009) Water and ecologically sustainable development in the courts. MqJICEL 6:129
Preston B (2013) Environmental public interest litigation: conditions for success. In: Conference presentation to the international symposium, Towards an effective guarantee of the green access: Japan's achievements and critical points from a global perspective, 30–31 March 2013, Awaji Island, Japan. Available http://www.lec.justice.nsw.gov.au/Documents/preston_environmental%20public%20interest%20litigation.pdf. 10 Oct 2016
Pring G, Pring C (2009) Greening justice: creating and improving environmental courts and tribunals. Access Initiative 48
Roberts A (2016) Second Reading Speech re Inclosed Lands, Crimes and Law Enforcement Legislation Amendment (Interference Bill), Hansard, NSW Parliament, 8 March 2016
Robinson J (2015) Environmental defenders tell inquiry that Coalition cuts will mean closure. The Guardian, 21 April 2015. Available https://www.theguardian.com/environment/2015/apr/21/environmental-defenders-tell-inquiry-that-coalition-cuts-will-mean-closures. 23 Nov 2016

Sarat A, Scheingold S (1998) Cause lawyering: political commitments and professional responsibilities. OUP, Oxford
Sullivan K (2016) Farmers protest Victorian Government water policy. The Weekly Times. 14 April 2016. Available http://www.weeklytimesnow.com.au/news/water/farmers-protest-victorian-government-water-policy/news-story/9a52320ec5f0408cb398cfc50855696b. Accessed 20 Nov 2016
Slezak M (2016) NSW anti-protest laws an attack on democracy, say legal experts. The Guardian, 15 Mar 2016
Thornton M (2016) The flexible cyborg: work life balance in legal practice. Sydney Law Rev 38 (1):1
Waschka M, Gardner A (2016) Diffuse source pollution and water quality law for the great barrier reef: why the reticence to regulate? In: Gray J, Holley C, Rayfuse R (eds) Tran-jurisdictional water law and governance. Routledge/Earthscan
Williams G, Kildea P (2010) The constitution and the management of water in Australia's rivers. Sydney Law Rev 32:595–616

Cases

Australian Conservation Foundation v Commonwealth (1980) 146 CLR 493.
Australian Conservation Foundation v Forestry Commission of Tasmania (1988) 19 FCR 127; 76 LGERA 369.
Australian Conservation Foundation v Minister for Resources (1989) 76 LGERA 200.
Boyce v Paddington Borough Council [1903] 1 Ch 109.
ICM Agriculture Pty Ltd v Commonwealth (2009) 240 CLR 140.
Fraser Island Defenders Organisation v *Hervey Bay Town Council* (1983) 2 Qld R 72; 51 LGRA 94.
Metgasco Ltd v Minister for Resources and Energy [2015] NSWSC 453.
Onus v Alcoa (Aust) Ltd (1981) 149 CLR 27.
Sinclair v Maryborough Mining Warden (1975) 132 CLR 473.
Tasmanian Wilderness Society Inc v Fraser (1982) 153 CLR 270 at 274.

Legislation

Crimes Act 1900 (NSW)
Environmental Planning and Assessment Act 1979 (NSW).
Environmental Protection and Biodiversity Conservation Act 1999 (Cth).
Water Act (Cth) 2007.
Water Management Act (NSW) 2000.

The Water Crisis in Flint, Michigan: Profitability, Cost-Effectiveness, and Depriving People of Water

Joseph W. Dellapenna

1 Introduction

At the end of September 2015, the press suddenly reported that people in Flint, Michigan were using water that contained heavy concentrations of lead for drinking, cooking, washing, and everything else (CBS News 2015). This was the first the general public heard of the problem (quickly recognized as a true crisis), but in fact there had been warnings about this and other problems with the city's water as early as April 2014 (Davis et al. 2016, p. 1). In August 2014, scientists at the US Environmental Protection Agency (EPA) and the Michigan Department of Environmental Quality identified elevated lead and e-coli levels in Flint's water. People were already complaining of a bad color, odor, and taste (Roy 2015). An epidemiological study by Dr. Mona Hanna-Attisha made public on 22 September 2015 (Davis et al., p. 21) was dismissed by the Michigan Department (along with the Michigan Department of Health and Human Services) as promoting hysteria (Davis et al. 2016, pp. 24–25, 28, 30–33). County officials flagged a spike in Legionellosis (Legionnaires' disease) in March 2015, but that was also dismissed by the Michigan departments (Davis et al. 2016, pp.1, 7, 17, 28, 33–34, 47–48; Goodnough 2016a). But finally, when a team of graduate students from Virginia Tech University under the leadership of Professor Marc Edwards published its findings, the press took notice and made the problem a public scandal (*id.*, pp. 20–21; Hohn 2016).

As a result of these problems, numerous children suffered the results of lead ingestion, including slowed intellectual development and other health effects (Centers for Disease Control 2016; Davis et al. 2016, pp. 23–24, 33–34, 58–60; Hanna-Attisha et al. 2016; Markowitz and Rosner 2013). Adults also showed health effects

J.W. Dellapenna (✉)
University of Houston Law Center, Houston, TX, USA
e-mail: dellapen@law.villanova.edu

L. Westra et al. (eds.), *The Role of Integrity in the Governance of the Commons*,
DOI 10.1007/978-3-319-54392-5_6

from the lead, while the spikes in Legionnaires disease (at least two spikes, lasting several months each) resulted in at least 12 deaths (Goodnough 2016a). Elevated e-coli levels also produced disease and probably deaths, although how many has not been established.

The question immediately arose: Who was to blame? Michigan's Republican governor, Rick Snyder, as well as Republicans in the US Congress sought to blame individual employees of the EPA and the Michigan Department of Environmental Quality as responsible for the problems in Flint (*e.g.*, Lee 2016). Was this tragedy just the result of individual malfeasance? Or was it the result of systemic failures caused by policy choices made at the highest levels of the state and federal governments? The answers to these questions have important consequences for how to respond to the tragedy. If primarily a question of individual malfeasance, then prosecution or at least dismissal is the appropriate response, not only as punishment but also as deterrence. If systemic failure, then different policies must be enacted and implemented. This chapter explores how the Flint crisis came about and what lessons American and other societies should draw from the event. I begin not with the water crisis, but with the evolution of the city of Flint.

2 Flint's Rise and Decline

Flint today is a derelict city. Its population on 1 July 2015 was 98,310, 57% African American, with non-Hispanic whites forming 36% of its population (US Census 2015). More importantly, Flint residents have high unemployment and had few economic prospects, even before the water crisis hit the press. Even the few signs of gentrification that began in 2004 were cut off by the city's fiscal crisis at the end of the decade (Scorsone and Bateson 2011). The bad press about the city's water did nothing to improve the economic prospects of the city's residents.

Flint was not always a derelict city. After its founding in 1819 (it was incorporated in 1855), it experienced slow but steady growth throughout the nineteenth century. Flint became an early center of the auto industry, playing a major role in the rise of Detroit (and Michigan more generally) as the global center of auto manufacturing. The Buick Motor Car Co. and the Chevrolet Co. were founded in Flint, as was the Fisher Body Co. that made auto bodies for these and other manufacturers. In 1908, Bill Durant, the president of Buick, merged the three companies (Buick, Chevrolet, and Fisher) to create General Motors Corp. While Durant moved the General Motors headquarters to the Fisher Building in uptown Detroit in 1928, a great deal of the company's manufacturing remained centered in Flint. Durant's personal wealth was wiped out in the stock market crash of 1929 and he lost control of the company to Alfred Sloan. He returned to Flint where he worked managing a bowling alley until his death in 1947 (Pelfrey 2006).

Flint also played a major role in the development of the American labor movement. In 1937, the United Auto Workers staged a "sit-down" strike in the General Motors factories in Flint. Workers occupied the plants for 44 days, successfully defying the police and the Michigan National Guard (the state's military force) (Fine 1969). Reluctantly, General Motors signed its first union contract to end the strike. With the high wages and benefits achieved through collective bargaining, workers in the four General Motors factories in Flint achieved middle-class lives and sent their children to college. By 1970, Flint was the second largest city in Michigan, with a population of about 200,000. The city was prosperous, middle-class, and mostly white.

Flint's decline began after 1970, when General Motors began moving jobs out of Flint to new factories in Mexico (Dandaneau 1996). Michael Moore started his career as a documentary filmmaker with his film "Roger and Me," about his efforts to obtain an interview with Roger Smith (then CEO of General Motors) about the transfer of jobs to Mexico, as well as portraying the increasingly dismal lives to the people left behind (Moore 1989). As people lost the opportunity for work in Flint, they began to move away, leaving behind people who, for one reason or another, were unemployable or who were too demoralized to seek work elsewhere.

Flint was now caught in the same sort of downward spiral that would lead to Detroit filing for bankruptcy in 2013 (Davey and Walsh 2013), several decades earlier and on a smaller scale. Flint, like all cities in such circumstances, faced an eroding tax base at the very time that the residents of the city need more investment (not less) in schools, infrastructure, business development, health care, and simple survival. Flint, starved for funds, began to run increasingly large fiscal deficits. With the state and national governments turning more and more to austerity budgets, there was little or no help available for such cities. Critics of such failing cities decried what they saw as "irresponsible liberals" under whose leadership the city refused to "live within its means" (*e.g.*, Reich 2013; Wilson 2016).

However one wants to characterize Flint's plight, the downwards spiral continues. The population continued to decline during the global economic crisis, and declined even more sharply as a result of the water crisis. The people left in Flint increasingly are trapped by poverty and now by the complete inability to sell homes in the city (Bosman 2016a). The challenge to society is how to rescue derelict cities like Flint, for Flint is hardly the only example of a city in sharp decline.

3 Rescuing a Derelict City in an Era of Deindustrialization

Michigan had a large and growing number of cities in similar situations as Flint. The state responded by enacting a law in 1988 authorizing the governor to appoint an "Emergency Financial Manager" (later renamed an "Emergency Manager") to take control of and administer cities facing insolvency (Michigan Compiled Laws 2012). The Emergency Manager's sole responsibility was to balance the city's books, by selling assets, repudiating contracts, slashing municipal salaries, and cutting city pensions (Collins 2015). Emergency Managers appointed in a number of Michigan cities over the next 28 years slashed salaries and cut pensions, repudiated long-term supply contracts for the city and collective bargaining agreements, but never repudiated a city's debts (Gillette 2014). The banks and other financial institutions would be protected even as the few decently paid jobs left in the city were to be paid less while being expected to take on more work as the total number of city employees were also reduced and persons living off pensions would be forced to tighten their belts. All of this would simply continue the downward economic spiral within a city blessed with an Emergency Manager.

Flint's first such blessing with an Emergency Manager came in 2002, and lasted for about a year (Longley 2011). This intervention, while repudiating contracts, slashing salaries, and cutting pensions, was relatively modest and probably contributed to a brief flurry of gentrification that began in 2004. That gentrification, wherever it would have led the city had it continued, ended with the economic crisis of 2008. The improved financial picture created by the first intervention was already turning downward before the global economic crisis, and that crisis pushed Flint into what almost could be called a death spiral.

Rick Snyder, elected governor in 2010, soon turned his attention to Flint. He appointed Michael Brown as Emergency Manager for the city in 2011, and replaced him eventually with Darnell Earley in 2013 (Ed Kurtz served briefly between the tenures of Brown and Earley). Earley's training and experience was in public administration, and he would go on to be the Emergency Manager of Detroit when that city filed for bankruptcy in 2013 (Davey and Walsh 2013). Later still, he would be the state appointed administrator for the bankrupt Detroit public schools (Jackson 2016). In each position, Earley acted strongly on his sole responsibility—balance the books without hurting major creditors (Doidge et al. 2015). In the case of Flint, a major step he took was to repudiate Flint's long-standing contract whereby it bought its public water supply from the city of Detroit (Fonger 2013).

4 Flint's Changes Its Water Supply

Most cities in the United States in the nineteenth century, beginning with Philadelphia in 1812, developed publicly funded water systems. By the mid-twentieth century all large and most smaller cities in the United States had municipal water systems that provided water at modest cost universally throughout the community (Troesken and Geddes 2003). Flint's first water delivery system was organized privately in 1888 and taken over by the city in 1903 (Davis et al., p. 15). Flint built its first water treatment plant in 1917, and a second in 1952 (Gringlas 2015). Blocked by the state from developing a new diversion system directly from Lake Huron, Flint found it economical to buy water for its municipal system from the Detroit Water Authority (then the Detroit Water and Sewage Department), signing a 30-year contract on 20 December 1965 (Davis et al., p. 15). The contract was continued thereafter on a year-by-year basis until it was repudiated at the behest of the emergency manager on 16 April 2013 (*id.*, p. 16). The Detroit Water Authority draws its water supply from Lake Huron, which is cleaner than waters from small rivers and lakes within Michigan. The Detroit Water Authority also treats the water for its customers, ranging from individual homeowners to neighboring municipalities like Flint.

The Emergency Manager determined that repudiating the contract with the Detroit Water Authority would save the city US $5,000,000 per year. While he forced Flint to face the need to save money on its municipal water service in a particularly acute way, this problem is hardly unique to Flint. Cities and towns all over the world have been forced to privatize their municipal water utilities in the hope of cutting costs and improving service. Caught between an ideological commitment that markets are the best way to resolve any problem and declining revenues caused not only by de-industrialization (as in Flint's case) but also by a growing anti-tax fervor, municipalities have seen private investors as the sole means to raise needed capital for investment in their water utilities. All too often, however, the result has been increased costs and deteriorating service (Borrowing Trouble 2013).

There has also been considerable public resistance to the privatization of public water systems (Herzog 2016). Demonstrations, riots, and elections have all been used, sometimes successfully, to prevent privatization. In extreme cases, cities have been forced to buy their water utilities back at considerable financial loss. The formulation of a "human right to water" was another step to turn back the tide of privatization. In Michigan, at least, Emergency Managers served to defeat such resistance. When Detroit (also under Emergency Manager Earley) was blocked in an attempt to privatize the city's water utility, Earley instructed the utility to cut costs and recover unpaid water bills in the hope of recovering some hundreds of thousands of dollars. Many impoverished home owners whose water bills were in arrears were threatened with a water shut-off if they didn't bring their bills up-to-date (Poupore 2015). Large commercial and industrial users (some of whose unpaid bills were in the millions) were not so threatened.

5 How Flint's Water System Failed

Flint did not privatize its water service, and it knew that to switch from the Detroit Water Authority would require investing in an upgraded treatment system. Flint planned to solve this problem by joining a consortium of communities in its region, to be called the Karegnondi Water Authority,[1] to establish a less expensive diversion from Lake Huron (Fonger 2010). During the interim between abandoning the connection to the Detroit Water Authority and the advent of the new Karegnondi Water Authority, Flint reactivated an old water withdrawal system with an aged water treatment plant (Davis et al., p. 16). Emergency Manager Earley decided that it was too expensive to do the necessary upgrades on the aged plant (*id*., pp. 7, 41),

[1]The name "Karegnondi" is from the Petan Indian name for Lake Huron.

setting the stage for the ensuing tragedy (*id.*, pp. 2, 18, 39–42). An employee of the Michigan Department of Environmental Quality warned that the switch would endanger Flint's water quality before the switch was made, but the Department advised Flint that it would be adequate to test the water every 6 months (*id.*, pp. 16–17, 26–30). On 13 and 14 October 2014, members of the governor's staff argued for a return to Detroit water and General Motors announced that it would no longer use water supplied by Flint in its remaining factory, both because of rising lead levels, but nothing was done to correct the problem (*id.*, pp. 17–18). Interestingly, the engineering firm that was in charge of the transition to Flint River water declined to answer questions put to it by the Flint Water Advisory Task Force in 2015 and 2016 (*id.*, p. 3 n. 4). That firm and another would later be sued by the state as bearing a major responsibility for the tragedy (Smith and Bosman 2016).

Given when Flint experienced its greatest growth, it was no surprise that the water mains as well as the service lines to the individual homes and businesses were 75 or more years old. Such old mains and lines were mostly lead pipes. Those lead pipes were critical to reducing waterborne diseases and were considered state-of-the-art at the time (Pease 2016). The pipes had, over the years that they carried water from the Detroit Water Authority, become encrusted with a protective coating of calcium carbonate that prevented the water from direct contact with the lead. The engineers monitoring the water as it passed through the treatment plant noted elevated levels of trihalomethanes that form when chlorine (added to disinfect the water) reacted with organic matter in the more polluted Flint River water. Chloride was then added to the water to the water to remove the organic matter at a ratio of nearly four times the level that had been necessary for the water supplied by the Detroit Water Authority. This greatly increased the corrosiveness of the water flowing through the pipes, quickly eroding the protective calcium carbonate (Roy 2015; Torrice 2016). Orthophosphates that might have prevented the leaching of lead into the water were not added, presumably to save on costs (Pease 2016).

The lead levels in the water supplied by the Detroit Water Authority had averaged 2.3 parts/billion (ppb), well below the 5 ppb that the EPA considers "cause for concern." The Virginia Tech study found that lead levels in 271 tested homes averaged 27 ppb (Lin and Park 2016). In Flint's eighth ward, the water in the homes averaged 158 ppb. The highest level the Virginia Tech study found was 13,000 ppb. The EPA considers a level of 5000 ppb to be a toxic waste. The amount of led in Flint's delivered water had been tested by the Michigan Department of Environmental Quality with a smaller sample of buildings. Had the Department including all samples in their averaging, by their own standards they would have had to intervene with decisive action earlier, but they excluded the most extreme of their samples as outliers that probably were not accurate. This made the test results acceptable (Davis et al., p. 18, 29; Keller and Watkins 2016). The publication of the Virginia Tech study on 27 August and 8 September 2015 forced officials' hands. Governor Rick Snyder first publicly acknowledged the problem at a press conference on 30 September, and on 8 October 2015, he proclaimed a water emergency in Flint. The state undertook to provide bottled water to all residents for all uses until corrective measures could be taken. Mid-level officials in the Michigan Department of Environmental Quality lost their jobs over their handling of the crisis (Associated Press 2015; Pérez-Peña 2016; Smith 2016b).

President Barak Obama did not proclaim a federal emergency until 16 January 2016, but the federal response was less central to the water crisis (Associated Press 2016a). Some lower level state officials (but no policy makers) were indicted, and civil litigation quickly multiplied (Associated Press 2016b; Haimerl and Goodnough 2016; Hall 2016; Macaluso 2016a; Raupe 2016; Smith and Bosman 2016). Legislative hearings were held at the state and federal level (*e.g.*, Goodnough and Steinhauer 2016). Yet even after the lead problem was acknowledge, state officials did not respond to the spiking Legionnaires disease. Governor Snyder did not acknowledge it until January 2016 (Goodnough 2016a). During the 10 months from the first noting of this problem by country health officials, there had been 87 cases and 12 deaths.

6 What Lessons Might Be Learned

Politicians from across the political spectrum quickly took sides on the Flint water crisis, with Republicans and Democrats reacting differently to the crisis and how to respond to it. Republicans tended to seek lower level officials to blame, seeing it as yet more evidence of how government cannot be relied upon to protect the public. Democrats tended to see wrongheaded policies enacted by a Republican governor and legislature as the root of the problem, along with steady reductions in funding for state and federal regulatory agencies enacted by Republican majorities in the state legislature and Congress. Republicans sought to deflect blame from Governor Snyder and from Republican policies (Bologna 2016b; Goodnough 2016b, c). They therefore pointed at scientists at the Chicago office of the EPA who had flagged the lead problem early in 2015, but had done nothing beyond sending a warning to the Michigan Department of Environmental Quality—where the warning was ignored (Davis et al., pp. 18–20, 48–53; Goodnough 2016b; Leven 2016b, c; Young 2016). This also enabled Republicans to blame "bureaucrats" in the Michigan departments for failing to do their jobs properly (Atkinson et al. 2016; Smith 2016a, b; Ware 2016b).

The EPA was in part responsible for what happened in Flint (Davis et al. 2016, pp. 8–9, 12, 48–53). The primary responsibility, however, lay with the state government (particularly the Michigan Department of Environmental Quality, but leading right back to the Governor who was in the loop from the beginning), as even the Flint Advisory Water Task Force appointed by Governor Snyder concluded (*id.*, pp. 2, 6–7, 26–30, 35–39, 40, 48–53). Despite loud protestations of Republicans about their determination to right the wrong that had been done, neither the Republican Congress nor the Republicans who dominated the Michigan legislature would fully fund the necessary remedial steps (Dabbs 2016a; Davenport 2016a; Goodnough and Steinhauer 2016). Curiously, the state legislature that professed that it didn't have time to consider additional funding did find time to outlaw anal sex (Rodriguez 2016). In a spasm of generosity, the state did eventually agree to pay the Flint residents' water bills, which they were still being charged for their undrinkable water (Bruninga 2016).

Democrats saw something more systemic at work, something that was rooted in policy choices made by governments that made the failures of individual officials predictable, failings that could have been avoided had different policy choices been made (Davis et al. 2016, pp. 10–11). First of all, while Flint's lead and other health problems in its water may have been more pronounced or dramatic than elsewhere, these problems are hardly unique to Flint. Such problems are distressing more widespread in the United States than generally acknowledged (Bologna 2016a; Cornwell et al. 2016; Leven 2016a; Olsen and Pullen 2016; Pappas 2016; Taylor 2016; Wines 2016). As with the Michigan Department of Environmental Quality, the responsible regulatory agencies (including the EPA) have tended to minimize or dismiss such problems, at least until the Flint crisis threw a spotlight on the problems.

It is tempting to consider the regulatory agencies' minimization or dismissal of poor water quality issues as simple examples of bureaucratic self-preservation at its most venal. Such minimizations and dismissals were examples of bureaucratic self-preservation, not of venality but of agencies being forced by repeated budget cuts to reduce the amount of work to be done to manageable levels (Adragna 2016a). The EPA and the Michigan Department of Environmental Quality (and, for that matter, the [federal] Centers for Disease Control), and municipalities like Flint) have been systematically starved of funds over more than two decades.[2] The very same congressmen who would lake seek to blame the EPA for the Flint water crisis had been, and continue to be, the congressmen who lead the fight to defund the EPA (Dabbs 2016b; Davenport 2016b, c). Indeed, they set about to claim that the failures of the EPA scientists in Chicago to take enforcement action when they first flagged the Flint problems demonstrates the inefficacy of governmental regulation in general—let markets solve the problem. One could describe this as a deliberate strategy of "starving the beast" until it was so small that it could be "drowned in a bathtub" (Norquist 2001). The Republicans in Congress continued to push reducing funds for the EPA water programs even after the Flint crisis (Adragna 2016b; Dabbs 2016b).

If the market proponents cannot sell the idea of letting the market provide personal water supplies to everyone (if the public water supply is unreliable, "let them drink bottled water" to paraphrase Marie Antoinette), they settle on the idea that the only proper way for governments to make decisions is through a cost/benefit analysis to determine how much to spend on protecting public water

---

[2]This is true at the federal level as well. As of 2016, the "The EPA has 200 fully authorized federal law enforcement agents who can carry firearms, 70 forensic scientists and technicians, and 45 attorneys who specialize in environmental crimes enforcement" (Coleman-Adebayo and Berends 2016).

supplies. Put another way, cost effectiveness is the proper test for public expenditures (Bosman 2016b). But what is the monetary value of safe drinking water? Governor Snyder himself dismissed the problem as involving "a few IQ points; it's not the end of the world" (Boggioni 2016). How much are a few IQ points worth? For how many people?

It would cost billions of dollars to replace all the led piping in Flint (Bosman 2016b; Ware 2016a). Supplying sufficient bottled water for Flint's declining number of residents for drinking, washing, cooking, and other purposes for the interval between the recognition of the problem and the time when the water to be delivered by the Karegnondi Water Authority could reestablish a sufficient coating of the pipes to bring lead levels within tolerable limits would cost a lot less. One effort to remedy the problem involved finding funds for Flint to return to water from the Detroit Water Authority (Davey 2015). That in itself would not resolve the problems because it would not immediately recreate the missing calcium carbonate linings in the pipes. Missing from such an analysis is any appreciation of the difficulties of living entirely dependent of bottled water. Is this a practical solution for washing clothes? How often would you wash your hands if you had to use a bottle for the necessary water? The latter problem has already manifested itself with a spiking of shigellosis, a disease spread by the infrequent washing of hands (Hauser 2016). No wonder the Flint Water Advisory Task Force ultimately recommended "Complete Lead Service Line Replacement" despite the cost that would entail (Davis et al. 2016, at 60–61).

7 Did Race Play a Role?

Governor Snyder has argued that there is no reason to believe race played a role in the development of the Flint water crisis (Pérez-Peña 2016). After all, the most important decisions were made by Emergency Manager Darnell Earley, and Earley is a black man. Furthermore, Flint's population is not entirely African-American; some 36% of the people in the city are non-Hispanic whites. Governor Snyder also claims that he acted decisively immediately upon discovering the problem. He claims that he knew nothing of the problems until 30 September 2015, and declared the water emergency of 8 October 2015.

The evidence supports the charge of environmental injustice, as even Snyder's own task force concluded (Davis et al. 2016, pp. 9, 13, 15, 54–56; Hammer 2016). Emails show that Snyder and his staff knew of the problems a full year before he took action (Boggioni 2016; Macaluso 2016b; Ridley 2016). Would Snyder have neglected the emerging problem for a year if the population of Flint still consisted mostly of middle-class whites? Nearly all the population of Flint are low-income or impoverished people, so perhaps it's more a case of class discrimination than of racial discrimination (Latimer 2016). Would Snyder have been so neglectful if Flint had included a sizable proportion of likely Republican voters?

For decades, Republican political campaigns have described public expenditures as a “give-away” to “those people” (Krugman 2016). When the average voter considered that public expenditures benefitted all in society, such expenditures had broad public support. As more and more people internalized the view that such expenditures benefit only “those people” (whom “those people” are was never explicitly stated), public support for public expenditures steadily eroded. This erosion enabled the election of anti-tax-and-spend politicians and even fueled a “tax-payers’ revolt” in the form of the TEA Party (“Taxed Enough Already” Party). Hence the policy of “starving the beast” is itself rooted in racism, or at least classism.

Finally, given the centrality of the Emergency Manager to the Flint crisis, it is worth noting that every city in Michigan that has been placed under an Emergency Manager has a black majority (Bosman and Davey 2016). That the Emergency Manager who made the critical decisions in Flint leading to the water crisis was himself black doesn’t preclude the possibility that the institution itself is racist. If the institution is solely devoted to protecting banks and other financial institutions at the expense of predominantly black populations, the racial classification of the particular Emergency Manager hardly seems relevant. It is perhaps noteworthy that in Flint as in other cities under an Emergency Manager, the one employment category exempted from the slashing of salaries were the police. Not only is police support necessary to enable the Emergency Manager to impose his decisions on the community, but police characteristically (even in predominantly black communities) are the whitest of employment categories (Ashkenas and Park 2015). No doubt just a coincidence that they were protected along with the banks and financial institutions.

8 Conclusions

One is tempted to ask whether the Flint water crisis was an investment crisis—a crisis brought on by underinvestment in water infrastructure? Or was it an equity crisis—a failure to treat the least among us as equally deserving of social and governmental solicitude? Or was it a democracy crisis—a failure to allow the people of Flint to govern themselves, imposed through the institution of the Emergency Manager (Davis et al. 2016, pp. 8, 11–12)? The answer, of course, is that it was all three (Pope 2016). The problem can be resolved only by addressing it on all three levels, particularly by introducing greater transparency in how governments (and businesses) conduct their affairs (Fishman 2016; Fournier 2016).

References

Adragna A (2016a) EPA's response to Flint crisis impeded by state: McCarthy. BNA Environ Rep 47:251

Adragna A (2016b) Bill would put environmental programs on chopping block. BNA Environ Rep 47:297

Ashkenas J, Park H (2015) The Race Gap in America's Police Departments. New York Times blog, 8 April. Available at http://www.nytimes.com/interactive/2014/09/03/us/the-race-gap-in-americas-police-departments.html

Associated Press (2015) Michigan: Regulator Quits in Drinking Water Crisis. New York Times, 30 December, at A13

Associated Press (2016a) Federal Emergency Is Declared in Flint over Contaminated Water. New York Times, 17 January, at A19

Associated Press (2016b) In Plea Deal, Ex-Michigan Official Admits Silence in Flint Water Crisis. New York Times, 15 September, at A15

Atkinson S, Haimeri A, Pérez-Peña R (2016) Anger and Scrutiny Grow over Poisoned Water in Michigan City. New York Times, 16 January, at A9

Boggioni T (2016) This Quote from Rick Snyder's Emails Says Everything You Need to Know about Flint's Water Crisis. Raw Story, 29 January. Available at http://www.rawstory.com/2016/01/this-quote-from-rick-snyders-emails-says-everything-you-need-to-know-about-flints-water-crisis/

Bologna MJ (2016a) Chicago exposed residents to lead-tainted water, suit says. BNA Environ Law Rep 47:655

Bologna MJ (2016b) Michigan AG says governor not cooperating in Flint probe. BNA Environ Law Rep 47:2010

Bosman J (2016a) Many Flint Residents Are Desperate to Leave, but See No Escape. New York Times, 5 February, at A12

Bosman J (2016b) Why It Won't Be Easy or Cheap to Fix Flint's Water Crisis. New York Times, 13 February, at A15

Bosman J, Davey M (2016) Water in Flint and Schools in Detroit Have Many Fuming over State Control. New York Times, 23 January, at A9

Borrowing Trouble: Water Privatization Is a False Solution for Municipal Budget Shortfalls (2013) Washington, DC: Food & Water Watch. Available at https://www.foodandwaterwatch.org/sites/default/files/Borrowing%20Trouble%20Report%20April%202013.pdf

Bruninga S (2016) Snyder signs $30 million Flint water assistance bill. BNA Environ Rep 47:713

CBS News (2015) Water Crisis in Flint Declared Public Health Emergency. 2 October. Available at http://www.cbsnews.com/news/water-crisis-in-flint-michigan-declared-public-health-emergency/

Centers for Disease Control (2016) Morbidity and Mortality Weekly Report: Blood Lead Levels among Children Aged <6 Years—Flint, Michigan, 2013–2016. 24 June

Coleman-Adebayo M, Berends K (2016) Water Crises Like Flint's Will Continue until the EPA Is Held Accountable. The Guardian, 9 Feb 2016. Available at https://www.theguardian.com/commentisfree/2016/feb/09/water-crises-like-flint-will-keep-happening-epa-accountable

Collins MB (2015) Pensions or paintings? Univ Miami Law Rev 14:1–29

Cornwell DA, Brown RA, Via SH (2016) National Survey of lead service line occurrence. J Am Water Works Assoc 108:E182–E191

Dabbs B (2016a) Republicans reject Flint amendments. BNA Environ Rep 47:842

Dabbs B (2016b) Republicans budget targets EPA 'activist' policies. BNA Environ Rep 47:855

Dandaneau SP (1996) A town abandoned: Flint Michigan confronts deindustrialization. SUNY Press, New York

Davenport C (2016a) Energy Bill Stalls as Sides Stand Firm on Flint Aid. New York Times, 5 February, at A17

Davenport C (2016b) Senate Approves Funding for Flint Water Crisis; House's Stance Is Unclear. New York Times, 16 September, at A18
Davenport C (2016c) E.P.A. Faces More Tasks, Louder Critics and a Shrinking Budget. New York Times, 19 March, at A11
Davey M (2015) Flint Will Return to Using Detroit's Water after Findings of Lead in Local Supply. New York Times, 9 October, at A16
Davey M, Walsh MW (2013) Billions in Debt, Detroit Tumbles into Insolvency. New York Times, 19 July, at A1
Davis MM, Kolb C, Reynolds L, Rothstein E, Sikkema K (2016) Flint Water Advisory Task Force Final Report. Available at www.michigan.gov/documents/snyder/FWATF_FINAL_REPORT_21March20 16_517805_7.pdf
Doidge M, Scarsone E, Taylor T, Sapotichne J, Rosebrook E, Kaminski D (2015) The Flint fiscal playbook: an assessment of the emergency management years (2011–2015). Michigan State University Extension, East Lansing, MI. Available at http://msue.anr.msu.edu/uploads/resources/pdfs/Flint-Fiscal-Playbook.pdf
Fine S (1969) Sit-down: the general motors strike of 1936–1937. University of Michigan Press, Ann Arbor
Fishman C (2016) The Water Data Drought. New York Times, 17 March, at A23
Fonger R (2010) Years in the Making, Karegnondi Water Authority Is Ready to Set a New Course for Water. Flint Journal, 23 October
Fonger R (2013) Flint Emergency Manager Endorses Water Pipeline, Final Decision Rests with State of Michigan. Flint Journal, 29 March
Fournier R (2016) How to Prevent the Next Flint. The Atlantic, 18 February. Available at http://www.theatlantic.com/politics/archive/2016/02/preventing-flint/463404/
Gillette CP (2014) Dictatorships for democracy: takeovers of financially failed cities. Columbia Law Rev 114:1373–1462
Goodnough A (2016a) Flint Outbreak Was Treated with Silence. New York Times, 23 February, at A1
Goodnough A (2016b) House Panel Denounces E.P.A. Actions in Flint Crisis. New York Times, 16 March, at A11
Goodnough A (2016c) Michigan Governor Tells Congress He Was Misled on Flint Water. New York Times, 18 March, at A1
Goodnough A, Steinhauer J (2016) As Grilling over Flint Water Begins, Partisan Divisions Surface. New York Times, 4 February, at A11
Gringlas S (2015) In Flint, Lead Contamination Spurs Fight for Clean Water. Michigan Daily, 3 December 2015
Hall N (2016) Litigating the Flint Water Crisis: Overview of the Lawsuits and Legal Action. Great Lakes Law blog. Available at http://www.greatlakeslaw.org/blog/flint-water-crisis/
Haimerl A, Goodnough A (2016) 6 More State Workers Charged over Flint Water Crisis. New York Times, 30 July 2016, at A14
Hammer PJ (2016) The Flint Water Crisis, KWA and Strategic Structural Racism. Wayne State University Law School Legal Studies Research Paper Series no. 2016–2017. Available at file:///C:/Users/jwdel/Downloads/SSRN-id2812171.pdf
Hanna-Attisha M, LaChance J, Sadler RC (2016) Elevated blood levels in children associated with Flint drinking water crisis: a spatial analysis of risk and public health response. Am J Public Health 106(2):283–290
Hauser C (2016) Bacterial Illness on the Rise in Flint as Residents Shun City Water. New York Times, 5 October, at A11
Herzog K (2016) No, Privatization Is Not the Way to Prevent the Next Flint. Grist, 19 February. Available at http://grist.org/living/no-privatization-is-not-the-way-to-prevent-the-next-flint/
Hohn D (2016) The ... Zealot. New York Times Magazine, 21 August, at MM40

Jackson A (2016) The ex-emergency manager of the ‘Flint water crisis’ Is Now the Emergency Manager of a School System in Crisis. Business Insider, 21 January. Available at http://www.businessinsider.com/darnell-earley-is-emergency-manager-detroit-public-schools-2016-1

Keller J, Watkins D (2016) How Officials Distorted Flint’s Water Testing. New York Times Blog, 29 July. Available at http://www.nytimes.com/interactive/2016/04/21/us/flint-lead-water-testing-distortions.html

Krugman P (2016) Michigan’s Great Stink. New York Times, 25 January, at A21

Latimer M (2016) Republicans Ignore a Poisoned City. New York Times, 21 January, at A27

Lee S (2016) EPA: Michigan officials failed in Flint crisis. BNA Environ Rep 47:541

Leven R (2016a) Between 15 million and 22 million people are served drinking water through lead pipes. BNA Environ Rep 47:768

Leven R (2016b) Congressmen call for resignations in Flint hearing. BNA Environ Rep 47:839

Leven R (2016c) Critics say that Flint complaint represents systemic EPA issues. BNA Environ Rep 47:2150

Lin JCF, Park H (2016) High Lead Levels Were Detected in Nearly 400 Flint Homes, and There May Be More. New York Times Interactive, 6 February, Available at http://www.nytimes.com/interactive/2016/02/04/us/flint-water-lead-maps.html

Longley K (2011) Flint would Be the only Michigan City to twice undergo state emergency takeover. 8 November. Available at http://www.mlive.com/news/flint/index.ssf/2011/11/flint_would_be_only_michigan_c.html

Macaluso N (2016a) Flint residents Sue over water contamination. BNA Environ Rep 47:250

Macaluso N (2016b) Michigan staff thought Flint water was City problem. BNA Environ Rep 47:267

Markowitz G, Rosner D (2013) Lead wars: the politics of science and the fate of America’s children. Universty of California Press, Berkely

Michigan Compiled Laws. 2012. § 141.1549

Moore M (1989) “Roger and Me.” Dog Eat Dog Films, distributed by Warner Bros

Norquist G (2001) As quoted on National Public Radio, Morning Edition, 25 May. Available at http://www.npr.org/templates/story/story.php?storyId=1123439

Olsen E, Pullen K (2016) What’s in your water? Flint and Beyond—analysis of EPA data reveals widespread lead crisis potentially affecting millions of Americans. National Resources Defense Council, Washington. Available at https://www.nrdc.org/sites/default/files/whats-in-your-water-flintbeyond-report.pdf

Pappas LA (2016) Pittsburgh testing finds high lead levels in water. BNA Environ Rep 47:2144

Pease CM (2016) Zika, lead, bad water, and why sex just Isn’t so fun nowadays. Environ Forum 33 (3):17

Pelfrey W (2006) Billy, Alfred, and general motors. AMACOM Publishing, New York

Pérez-Peña R (2016) Governor Denies Race Affected Flint Response. New York Times, 23 January, at A10

Pope C (2016) Three Reasons Flint’s Water Is Poisoned. Ecowatch, 16 February, Available at http://www.ecowatch.com/3-reasons-flints-water-is-poisoned-1882175019.html

Poupore M (2015) Turning off the tap: the right to water in the United States. Human Rights Brief 22(1):10

Raupe B (2016) Lawsuit seeks lead service line replacement in Flint. BNA Environ Rep 47:319

Reich R (2013) The myth of living beyond our means. 24 January, Available at http://robertreich.org/post/41395861898

Ridley G (2016) Snyder Ordered DEQ to Withhold Flint Lead Test Results, Emails Claim. 10 February. Available at http://www.mlive.com/news/flint/index.ssf/2016/02/snyder_ordered_deq_to_withhold.html

Rodriguez M (2016) Michigan Is too Busy Trying to Outlaw Anal Sex to Address Toxic Flint Water. Available at https://mic.com/articles/134600/michigan-is-too-busy-trying-to-outlaw-anal-sex-to-address-toxic-flint-water#.SWqALBe7F

Roy S (2015) The Unintended Consequences of Migrating to Flint River Water. 23 August. Available at http://flintwaterstudy.org/2015/08/the-unintended-consequences-of-migrating-to-flint-river-water/

Scorsone E, Bateson N (2011) Long-term crisis and systemic failure: taking the fiscal distress of America's older cities seriously: case study, Flint Michigan. Michigan State University Extension, East Lansing. *Available at* https://www.cityofflint.com/wp-content/uploads/Reports/MSUE_FlintStudy2011.pdf

Smith M (2016a) Flint Wants Safe Water, and Someone to Answer for Its Crisis. New York Times, 10 January, at A16

Smith M (2016b) Michigan Water Regulator Fired for Role in Flint Crisis as Fallout Continues. New York Times, 5 February, at A10

Smith M, Bosman J (2016) Contractors Are Blamed by State Over Flint. New York Times, 23 June, at A17

Taylor K (2016) Lead Poisoning Persists Despite Decades-Old Fight. New York Times, 1 September, at A21

Torrice M (2016) How lead ended up in Flint's tap water. Chem Eng News 94(7):26–29. Available at http://cen.acs.org/articles/94/i7/Lead-Ended-Flints-Tap-Water.html%20

Troesken W, Geddes R (2003) Municipalizing American Waterworks, 1897–1915. J Law Econ Org 19:373–400

US Census (2015) Quick Facts 2015: Flint city, Michigan. Available at http://www.census.gov/quickfacts/table/PST045215/2629000,00

Ware P (2016a) Flint water crisis offers lessons on switching sources. BNA Environ Rep 47:320

Ware P (2016b) Snyder criticized for hiring PR firm to handle flint. BNA Environ Rep 47:339

Wilson K (2016) Living beyond Our Means. McDonald County Press, 29 September. Available at https://www.pressreader.com/

Wines M (2016) Beyond Flint, lead poisoning persists despite decades-old fight

Young E (2016) Michigan governor backs emergency managers despite crisis. BNA Environ Rep 47:1072

International Law and Human Security: The Environmental and Geopolitical Impacts of China's Artificial Island-Building at Fiery Cross Reef

Kathryn Anne Gwiazdon

听他的话, 看他的行为 – Chinese proverb, "Listen to his words, watch his deeds." (Chan 2016, p. 65).

1 Introduction

At times, the wide and varied geopolitical issues of the South China Sea seem to rival its status as one of the most biodiverse areas on the planet (McManus 2017, p. 10; Trowbridge 2016). Here, the routine is not abnormal: another day, another coral reef of invaluable worth to the sea, to the fisheries, to the economies, to the people, is severely damaged or permanently destroyed (PCA 2016, p. 476; PCA 2016, para. 904, citing Merits Hearing Tr. (Day 4), p. 150, citing Prof. Carpenter; PCA 2016, para. 965, citing Lee, Ferse Report, and McManus Report). Another day, another fishing vessel is boarded, wherein endangered species are discovered purposefully harvested (Cochrane 2016; Lee 2016; PCA 2016, para. 950–964, and section D(2)(b)). Another day, another island is built (see generally, PCA 2016). Another day, another warship travels through the waters in a Freedom of Navigation operation (USDOS 2016). Another day, another warning that this is considered "an intentional provocation" (Callahan 2016). Another day, another research drone seized, creating an international incident (AFP 2016). Another day, another

Kathryn Anne Gwiazdon, J.D., Esq. is the Executive Director of the Center for Environmental Ethics and Law, a non-profit organization that seeks to highlight and promote environmental ethics in law and life. Her scholarly research focuses on comparative international environmental law and applied ethics.

K.A. Gwiazdon (✉)
Center for Environmental Ethics and Law, Charlottesville, VA, USA
e-mail: kathryngwiazdon@environmentalethicsandlaw.org

L. Westra et al. (eds.), *The Role of Integrity in the Governance of the Commons*,
DOI 10.1007/978-3-319-54392-5_7

weapons system is placed on an artificial island (Bodeen 2016). Another day, another nuclear-capable bomber flies overhead (Sharman 2016).

This article seeks to shed light on the international law and human security considerations of the construction, operation, and use of artificial islands by the People's Republic of China (PRC) in the South China Sea, and the environmental and geopolitical repercussions of these actions. It will do this largely through the analysis of the July 12, 2016 decision of the Permanent Court of Arbitration (PCA), *Philippines v. China*, a case regarding maritime rights and entitlements, the legal status of certain features, and the lawfulness of certain actions of the PRC under the 1982 UN Convention on the Law of the Seas (UNCLOS) (PCA 2016, para. 2).

The reason for the maritime claims is simple, yet circular: access to resources, and expansion of military power to protect those resources. Although there are maritime claims and development activities by several nation-states throughout the South China Sea, no other nation can compare to the speed, scale, and scope of the PRC (Lee 2016; USDOD 2015). Due to the breadth of the applicable subject-matter, this chapter will focus particularly on the PRC's island-building activities at Fiery Cross Reef, although it will look at the implications of the entire policy. Fiery Cross Reef is found in the Spratly Islands of the South China Sea, an area contested by the Nation of Brunei (Brunei), Malaysia, the Republic of the Philippines (Philippines), the PRC, Taiwan, and the Socialist Republic of Vietnam (Vietnam) (CIA 2016). It was once only a single uninhabitable rock visible at high-tide, yet now, it has become an island housing people and farm animals, with an airstrip and military capabilities (Mollman 2016; PCA 2016, para. 564).

This article will begin by setting the context of the region through a human security lens. It will look at the components that foster human security and how they correlate to the relationships and resources in the South China Sea. It will show how the area has the potential to pivot from one of the most hotly contested areas in the world, where tensions between states seem only to be increasing, to an area that, if properly and cooperatively managed, provides long-term health and security for the entire region.

It will then provide an overview of the artificial island-building actions of the PRC in the South China Sea, with a focus on the transition of Fiery Cross Reef into Yongshu Island. It will give an overview of the PCA decision and the environmental consequences of the PRC's conduct, to show how the transition from a reef to an island has immediate and long-term, sometimes permanent, environmental impacts for the region and the planet. It will then explore the far-reaching geopolitical consequences of such developments, and consider how other states are responding to the PRC's efforts to solidify its maritime claims, expand its access to resources, and widen its military reach.

Lastly, the article will explore some potential ways forward for the nations and neighbors of the South China Sea that promote environmental conservation and human security for the entire region. It will show that when the integrity of the ecosystems of the region is harmed, the integrity of the region, and every inextricably-linked species and state, is also harmed. Therefore, regional cooperation and conservation promotes human security, which in turn promotes national and regional security.

2 Fostering Human Security in the South China Sea

"The world can never be at peace unless people have security in their daily lives." (UNDP 1994). One of the most important documents to be written exploring human security, its components, and its operationalization is the 1994 United Nations (UN) Human Development Report (UNHDR 1994); (UNDP 1994). The 1994 UNHDR was drafted two years after the UN Conference on Environment and Development, which led to Agenda 21, the Rio Declaration on Environment and Development, the Statement of Forest Principles, the UN Framework Convention on Climate Change and the UN Convention on Biological Diversity (UNCED 2016). These landmark environmental instruments laid the groundwork for the 1994 UNHDR: human security cannot be separated from environmental security.

Human security, at its most basic sense, is the freedom from want, from violence, or from the fear of violence (UNGA 2012; see generally UNTFHS 2016). It is international society's consideration of the protection and empowerment of the rights of individuals, including future generations, in the face of potential threats from other individuals or nation-states. It is people-centered and looks at the daily lives of individuals and whether they live in conflict or in peace. For example, if the basis of one's livelihood is fishing and that way of life is threatened, human insecurity increases, as does the potential for conflict.

Human security is also comprehensive, multi-disciplinary, and context-specific (UNGA 2012; UNTFHS 2016). It includes development, international relations, and human rights (UNDP 1994; UNTFHS 2016). It is also a universal concern, so when human security is threatened, actors globally are likely to be involved (UNDP 1994; UNTFHS 2016). The effects from pollution or environmental degradation, or threats from, and the aftermath of, conflict, for example, are rarely confined within national boundaries.

To foster human security, it is important to address the root causes of threats and seek to prevent those threats (UNDHR 1994; see generally UNTFHS 2016). So, what causes conflict, and what prevents conflict? In the geopolitical arena, causes of conflict largely relate to territory and resources, and have a profound, direct regional and global impact on human security. For example, many nations that border the South China Sea are heavily reliant on the fisheries and habitats of the South China Sea (independent from the global demands on the region). It is a region that produces more fish than almost anywhere else on earth (Bale 2016). However, it is being severely over-fished and is near collapse (Bale 2016). If that food system collapses, what happens to the people reliant on that system, in terms of available food and economy?

2.1 Food Security

Food security means that all people, at all times, have physical and economic access to food (UNDP 1994). The world has enough food to go around, but not everyone has access to it, whether by growing it for themselves, or having the purchasing power to acquire it. Mass, and, as will be shown below, illegal, commercial fishing is having a profound impact on the daily lives and economies of people in the South China Sea region (Bohane 2016; McManus 2017, p. 4). Due to the contested nature of the waters, where multiple states hold conflicting claims, implementation of international treaties that would protect these waters is difficult, which results in great harm to the re-generational capabilities of the fisheries (McManus 2017, p. 4). Or, as shown in *Philippines v. China*, even when national laws do exist, they are arguably not being enforced (Bartley 2016; PCA 2016, para. 965).

Commercial and traditional fishermen are at the front line of the disputes in the South China Sea (Denyer 2016). There are frequent conflicts, often violent, between PRC military and fishing vessels and other nations, such as Indonesia, Japan, Malaysia, the Philippines, the Republic of Korea, and Vietnam (Bartley 2016; Bohane 2016). The risks of traveling in these waters is high, but the rewards are some of the foundations of life: fish, food, and economy.

In efforts to solidify maritime claims, the PRC is encouraging broader and increased commercial fishing, using fishermen as "important tools in expanding China's presence and the country's claims in the disputed waters." (Denyer 2016; McManus, p. 7–8). As one PRC fisherman noted, "It is our water, but if we don't fish there, how can we claim it is our territory?" (Denyer 2016).

Many individuals and communities in the region also rely upon traditional, or "artisanal", fishing (Bohane 2016; PCA 2016, para. 761, 794–796). Under international law, traditional fishermen have different rights across maritime zones (UNCLOS 1982, Art. 51(1), 62(3)). However, many of these fishermen can no longer access fishing grounds due to the military presence of contesting states. In particular, the conduct of the PRC "ha[s] created a deep sense of fear among Filipino fishermen that has significantly curtailed their fishing activities and severely impacted their ability to earn a livelihood." (PCA 2016, para. 680, 767–770, 810).

2.2 Environmental Security

Environmental security acknowledges humanity's complete and utter dependence on a healthy physical environment (UNDHR 1994). This includes non-contaminated soil to grow food, clean waters to drink and fish and bathe in, a program for sanitation, and non-polluted air to breathe. Water is life in the world, and this is no less so in the South China Sea. The entire region is incredibly dependent on healthy marine ecosystems to provide their food and economy.

Beyond human need, however, is the integrity of the natural systems for their own sake. Fostering such ecological integrity, however, can only foster human integrity, which in turn fosters human security (Westra 1994; Kintzele 2011).

The South China Sea is one of the most biodiverse regions in the world in terms of its fisheries and coral reef ecosystems, roughly five to ten times more diverse than Hawaii or the Caribbean (McManus 2017, p. 10). And portions of the South China Sea are also included in the Coral Triangle, a region of "anomalously high marine species diversity" that is the focus of a major conservation effort known as the Coral Triangle Initiative, which aims to conserve reefs and alleviate coastal poverty (McManus 2017, p. 10).

Coral reefs are some of the most biodiverse ecosystems on earth and are crucial for socio-economic development (PCA 2016, para. 823–824, citing McManus Report, Ferse Report). However, they are also incredibly fragile under human pressure (PCA 2016, para. 824, citing Ferse Report, Mora Report). The island-building is causing long term and permanent damage, displacing and destroying entire ecosystems, reefs, and species (PCA 2016, para. 1178–1179). Overfishing in the region is leading to lower and lower catches, and the inability of parent fishes to reproduce (McManus 2017, p. 3). In addition, vulnerable and endangered species are being targeted (McManus 2017, p. 4).

It is also becoming increasingly difficult for scientists to conduct research (McManus 2017, p. 9). From what is known, however, it is reasonable to infer that there could be unknown species, valuable for their own sake, as well as for the development of human health and science (McManus 2017, p. 9).

2.3 *Physical Security*

"Perhaps no other aspect of human security is so vital for people as their security from physical violence." (UNDP 1994).

Physical security is freedom from the threat of violence to one's self, either from the state or other individuals, groups, or states (UNDP 1994). Even though the PRC stated in 2015 that they do not intend to pursue militarization, militarization in the South China Sea is occurring (Bodeen 2016). In 2014, the PRC Army also held a sovereignty oath ceremony at James Shoal, well within Malaysia's EEZ (Wee 2014). In addition, and explained further below, PRC fishing vessels are protected by PRC military vessels (PCA 2016, para. 964, citing Chinese Ministry of Foreign Affairs). Satellite images from November 2016 have also showed that the PRC placed anti-aircraft and anti-missile weapons on all seven of their newly created artificial islands, including Fiery Cross Reef (Bodeen 2016). In addition, Vietnam has been extending their airstrips to account for military aircraft, and have fortified five islands with rocket launchers, all within striking distance to the PRC's islands (Bohane 2016). Both nations have claimed a right to self-defense (Bodeen 2016; Bohane 2016). Even the FON operations by US military vessels account for an overall militarization of the region (USDOS 2016).

These actions may appear localized, but they have global repercussions. Targeted at the PRC, the Group of 7 (G7) has made it clear that "countries should stop land reclamation and militarization in one of the world's busiest shipping lanes." (Lim 2016). China dismissed the statements as "hype." (Lim 2016).

In addition to the regional countries' maritime claims, the United States declared a "pivot" to the Asian Pacific region, with a goal to strengthen relationships with allies in the region, and have a more "flexible and sustainable" troop presence (Manyin et al. 2012, citing USDOD 2012). Underlying the pivot was a realignment of U.S. foreign policy, national security, and economic interests to the region (Manyin et al. 2012, citing USDOD 2012). In particular, the United States aimed to address China's aggressive island-building and militarization in the South China Sea, with particular attention to its implications for Freedom of Navigation (Manyin et al. 2012, citing USDOD 2012).

The other components that the UN considers when advancing human security include health security, community security, and political security. For purposes of this chapter, these will not be unpacked. What is important to remember is that these seven components of human security are interrelated, and "a threat to one element of human security is likely to travel, like an angry typhoon, to all forms of human security." (UNDP 1994). So, what happens when this typhoon is unleashed, when a nation undermines the human security of an entire region by (1) building islands; (2) on delicate reef ecosystems; and (3) in contested waters?

3 China's Artificial Island-Building: From Fiery Cross Reef to Yongshu Island

"South China Sea, our beautiful motherland, we won't let go an inch." (Bodeen 2016, citing a billboard in Weifang, PRC).

3.1 Philippines v. China*: The Policy of the PRC under International Law*

"The law of the sea is as old as nations, and the modern law of the sea is virtually as old as modern international law." (Henkin 1979).

On January 22, 2013, pursuant to compulsory jurisdiction as detailed in Part XV, Annex VII of UNCLOS, the Philippines commenced its action against the PRC with the Permanent Court of Arbitration (PCA) (PCA 2016, para. 4). Established in 1899, the PCA is an intergovernmental organization that serves as a dispute resolution mechanism for its members (PCA-CPA 2016). Although the PRC disputed the jurisdiction, proceedings, and award of the Tribunal, it is a binding decision under UNCLOS (UNCLOS 1982, Art. 296(1) and Annex VII, Art. 11).

The Philippines and the PRC are member states of the PCA, and both states are parties to UNCLOS.

The Philippines asked the Tribunal to resolve four matters: (1) that the PRC's rights and entitlements in the South China Sea must be based on UNCLOS and not on any claim to historic rights, i.e. that the PRC's nine-dash line demarcation is without legal effect; (2) the entitlements to maritime zones claimed by both the Philippines and the PRC, and that none of features claimed by the PRC (such as the artificial islands) generate an entitlement to an exclusive economic zone or to a continental shelf; (3) a series of disputes regarding PRC actions (including artificial island building) that infringe on the rights of the Philippines under UNCLOS, including preventing the Philippine's right to fish, failing to protect and preserve the marine environment by tolerating and supporting illegal fishing practices of PRC fishermen, and inflicting severe harm on the marine environment by building artificial islands; and (4) that the PRC has aggravated the disputes by continuing to build islands and restrict access to the waters during the course of the arbitration proceedings (PCA 2016, para. 7–10).

The PRC did not participate in the proceedings, and consistently rejected the authority and jurisdiction of the Tribunal, including its decision on 12 July 2016 (PCA 2016, para. 11; Phillips et al. 2016). UNCLOS Annex VII, however, specifically provides for the non-participation of a party in Article 9, the "[a]bsence of a party or failure of a party to defend its case shall not constitute a bar to the proceedings." (PCA 2016, para. 11). The PRC is still considered a party to the proceedings, and is bound by the determination of the Tribunal (PCA 2016, para. 12, citing UNCLOS 1982, Art. 296(1) and Annex VII, Art. 11, para. 143). Due to the absence of a Party, and as directed by UNCLOS, the Tribunal took extra measures to ensure fairness and efficiency (PCA 2016, para. 144).

The PRC did send several documents to the Tribunal to clarify their positions, but made note that the submissions should not be interpreted as participation in the proceedings (PCA 2016, para. 13). The Tribunal bifurcated the jurisdiction and merits; they held a separate Hearing on Jurisdiction in July 2015, and ultimately awarded Jurisdiction in October 2015 (PCA 2016, E, and para. 14, 60). The PRC responded that the arbitration will "lead to nothing":

> China has indisputable sovereignty over the South China Sea Islands and the adjacent waters. China's sovereignty and relevant rights in the South China Sea, formed in the long historical course, are upheld by successive Chinese governments, reaffirmed by China's domestic laws on many occasions, and protected under international law including the United Nations Convention on the Law of the Sea (UNCLOS). With regard to the issues of territorial sovereignty and maritime rights and interests, China will not accept any solution imposed on it or any unilateral resort to a third-party dispute settlement. (PCA 2016, para. 61, citing PRC Ministry of Foreign Affairs).

The Hearing on the Merits was then heard 24–30 November 2015 (PCA 2016, para. 14).

UNCLOS is a binding international treaty that fosters communication between nations and defines their rights and responsibilities with respect to the peaceful and equitable use of the world's oceans, and the "study, protection, and preservation of the marine environment." (UNCLOS 1982; Preamble; PCA 2016, para. 4). Known as "the constitution of the oceans," it was adopted to "settle all issues relating to the law of the sea," and has been ratified by 168 parties (PCA 2016, para. 4). It is considered to be, "the most extensive and detailed product of codification activity States have ever attempted and successfully concluded under the aegis of the United Nations.":

> The Convention has 320 articles, set out in seventeen parts, as well as nine annexes. Parts II to XI concern the different maritime zones: territorial sea and contiguous zone, straits used for international navigation, archipelagic waters, the exclusive economic zone, the continental shelf, the high seas, the International Seabed Area, and special provisions on the regime of islands and of enclosed and semi-enclosed seas. Parts XII to XIV concern specific marine activities and questions in all areas: the protection of the environment, marine scientific research, and the development and transfer of marine technology. Part XV (and Annexes 5 to 8) concerns the settlement of disputes. Parts XVI and XVII set out general and final clauses. (Treves 2008).

Much of the treaty is also enforceable as customary international law, so it is considered binding law to non-parties, such as the United States. Customary international law is a binding source of international law that results from the "customary practice of states followed from a sense of legal obligation." (Henkin 1987; see also Statute of the Intl. Ct. of Justice, Art. 38(1)(b)).

Also pertinent to this article is the right to innocent passage, and Freedom of Navigation (FON) operations; the former acknowledges the sovereignty of the coastal state, whereas the latter directly contests any claimed sovereignty (UNCLOS 1982, Article 17; Glaser 2015; USDOS 2016). Innocent passage is allowed in the territorial sea of a state (UNCLOS 1982, Article 17). Under innocent passage, a state may pass through waters in an expeditious and continuous manner, which is not "prejudicial to the peace, good order or the security" of the coastal state (UNCLOS 1982, Article 19(1)). Fishing, polluting, weapons practice, and spying are not considered "innocent." (UNCLOS 1982, Article 19(2)). FON operations allow states to challenge other nation's claims in international waters and airspace (Glaser 2015).

Fiery Cross Reef is one of several hundred reefs found within the Spratly Islands of the South China Sea. Vietnam, the PRC, and Taiwan each claim ownership of the entire set of islands, whereas Malaysia and the Philippines each claim partial ownership (CIA 2016; see also Rosenberg 2016). The PRC is not the only nation building in the South China Sea, but they are building the most, with the most amount of damage (McManus 2017, pp. 23–24). In 2015, the U.S. Pentagon estimated the total amounts of reclaimed land: China (2900 acres), Vietnam (80 acres), Malaysia (70 acres), the Philippines (14 acres) and Taiwan (8 acres) (USDOD 2015). The scale and scope of development, destruction, and long-term impact is simply incomparable.

Vietnam claims that they are the rightful owners of the Paracels and the Spratlys due to a King dispatching soldiers to the area in 1836 (Bohane 2016). They argue, "Sovereignty cannot be claimed by fishermen, but it can be claimed through edicts of a king." (Bohane 2016). The PRC claims it has "indisputable" sovereign rights due to historical records (PCA 2016, para. 61, citing PRC Ministry of Foreign Affairs). The PCA stated that customary international law dictates that historical claims cannot conflict with provisions of the Convention, and that historical fishing may be one element considered when determining maritime rights (PCA 2016, para. 238(b), para. 242, citing Art. 62).

The PRC's demarcation, known as the "nine-dash line", encompasses nearly the entire South China Sea, in stark contrast to the baseline demarcations as defined by UNCLOS and as claimed by their neighbors (Rosenberg 2016; UNCLOS 1982, Sec. 2; see also PCA 2016, p. 77 for map submitted to the UN by the PRC). UNCLOS does not make its delineations dependent upon historical markers, as different states may have different historical records, but rather on baseline rules applied to all states, with a noted exception in Art. 298 (UNCLOS 1982, Sec. 2, and Art. 298).

There have been numerous calls on the PRC to clarify what it legally asserts with its nine-dash line, which was originally an 11-dash line. The PRC sees these requests as an attempt to "intentionally confuse territorial disputes with disputes over maritime delimitation, deny China's sovereignty over the South China Sea Islands and their adjacent waters, and cover up their illegal invasion and occupation..." (PCA para. 270, citing the PRC Ministry of Foreign Affairs). Some PRC scholars view any clarification as an opportunity for foreign intervention; for example, the President of the China Institute of South China Seas Studies, Wu Shicun, states, "it may help remove U.S. concerns over the line's potential harm to U.S. domination in the South China Sea while minimizing China's growing capacity to defend its rights in the future, thus eliminating any legal hurdles to U.S. "freedom of navigation" in the South China Sea." (Shicun 2014). Philippines Supreme Court Justice Antonio Carpio, on the other hand, states, "China's nine-dash line claim simply cannot co-exist with UNCLOS – one kills the other," and that, "To uphold China's nine-dash line claim is to wipe out centuries of progressing the law of the sea." (Malig 2013).

The PCA noted this, "China has never expressly clarified the nature or scope of its claimed historic rights. Nor has it ever clarified its understanding of the meaning of the 'nine-dash line'." (PCA para. 180). The Tribunal ultimately ruled that:

> as between the Philippines and China, China's claims to historic rights, or other sovereign rights or jurisdiction, with respect to the maritime areas of the South China Sea encompassed by the relevant part of the 'nine-dash line' are contrary to the Convention and without lawful effect to the extent that they exceed the geographic and substantive limits of China's maritime entitlements under the Convention; and further DECLARES that the Convention superseded any historic rights, or other sovereign rights or jurisdiction, in excess of the limits imposed therein; (PCA 2016, p. 473).

And that Fiery Cross Reef, in particular, "in [its] natural condition, are rocks that cannot sustain human habitation or economic life of their own." Therefore, there is no entitlement to an exclusive economic zone or continental shelf (PCA 2016, para. 474). Under Article 121(3) of UNCLOS, Fiery Cross Reef is considered a high tide water feature, and that prior to the PRC's construction, it was a single rock visible at high tide (PCA 2016, para. 564). The construction of the artificial island did not change its designation under international law as a high tide water feature, "China's construction on Fiery Cross Reef, however extensive, cannot elevate its status from rock to fully entitled island." (PCA 2016, para. 565).

3.2 Environmental Impacts

A major component of the Tribunal's decision considered the environmental impacts of PRC's actions in the South China Sea, around and including the artificial islands (PCA 2016, para. 9, VII, VIII). The Philippines argued that the PRC breached Part XII of the Convention (Articles 192–237), Protection and Preservation of the Marine Environment, and particularly Article 192 (general obligation to protect and preserve the marine environment) (PCA 2016, para. 907). The Philippines noted that the interpretation of Article 192 may also help guide obligations under Convention on the Illegal Trade of Endangered Species (CITES) and the Convention on Biological Diversity (CBD):

> (1) To protect and preserve marine ecosystems: The Philippines observes that coral reefs are a fragile and vitally important part of the marine ecosystem and argues that "creating artificial islands out of coral reefs is the worst possible way to treat these fundamental ecological building blocks."
>
> (2) To ensure sustainable use of biological resources: The Philippines notes that this reflects a long-recognized duty to conserve living resources. Blast fishing and the use of cyanide [by China] are wasteful and unsustainable, and violate Articles 192 and 194 (including as marine pollution) whether the methods are used in the territorial sea or beyond.
>
> (3) To protect and preserve endangered species . . .
>
> (4) To apply a precautionary approach in all of these respects. . ..
>
> (5) To consult and cooperate with the relevant coastal States: . . .[due to the] "characteristic regional features." . . .very little evidence of genuine Chinese cooperation... The Philippines considers China's behavior towards the Philippines and other States. . . to be aggressive rather than cooperative. (PCA 2016, para. 908, 910).

The Tribunal considered it well established that Article 192 imposes a "general obligation" that extends both to "protection" of the marine environment from future damage and "preservation" in the sense of maintaining or improving its present condition (PCA 2016, para. 941, citing *Legality of the Threat of Use of Nuclear Weapons*, Advisory Opinion, ICJ Reports 1996, p. 226 at pp. 240–242, para. 29). Therefore, the PRC has "a positive duty to prevent, or at least mitigate significant harm to the environment when pursuing large-scale construction activities." (PCA 2016, para. 941, citing Indus Waters Kishenganga Arbitration (*Pakistan v. India*), Partial Award, 18 February 2013, PCA Award Series (2014), para. 451; quoting Arbitration Regarding the Iron Rhine ("IJzeren Rijn") Railway between the Kingdom of Belgium and the Kingdom of the Netherlands, Award of 24 May 2005, PCA Award Series (2007), RIAA Vol. XXVII p. 35 at pp. 66–67, para. 59).

The Award ultimately found that that the PRC violated Article 192 for harvesting endangered species; for the severe, irreparable harm to the coral reef ecosystem; and for their lack of cooperation and communication with neighbors concerning the protection of the marine environment (PCA 2016, p. 475). The Tribunal also found that the PRC aggravated the dispute by continuing to cause extensive and irreparable damage during the proceedings of the Tribunal, including permanently destroying the natural evidence at the reef (PCA 2016, p. 476).

In 1997, then Vice-President of the International Court of Justice, Christopher Weeramantry, wrote a Separate Opinion on the concept of sustainable development in international law in which he determined that "environmental rights are human rights", as "the ethical and human rights related aspects of environmental law bring it within the category of law so essential to human welfare." (Weeramantry 1997, p. 111). He concluded:

> Sustainable development is thus not merely a principle of modern international law. It is one of the most ancient of ideas in the human heritage. Fortified by the rich insights that can be gained from millennia of human experience, it has an important part to play in the service of international law. (Weeramantry 1992, p. 107-108).

All nations, including the PRC, are ethically and legally obligated to act in good faith to prevent irreversible damage to the environment (Kintzele 2011). Yet, the environmental damage at Fiery Cross Reef is so extensive that some scholars compare it to an environmental weapon of mass destruction:

> ...the long-term damage done is indiscriminate and undirected... In attempting to exclusively secure and control natural resources, [China] is also destroying the most fragile and sensitive marine resource base of the South China Sea and diminishing everyone else's. It has engaged in environmental aggression on a regional scale, and turned reclamation into an environmental weapon of mass destruction. (Batongbacal 2015).

In order to build the islands, large dredgers destroy and vacuum nearby reefs and sand, and pump the debris directly onto the reef (PCA 2016, para. 978–979 citing Ferse Report; McManus 2017, p. 17). Therefore, not only do the reefs below the islands "disappear entirely", but nearby reefs are also damaged (PCA 2016, para. 978, 982, citing Ferse Report). The dredging is also used to create deep shipping channels, permanently altering the sea structure to allow for increased traffic (PCA 2016, para. 982, citing Ferse Report). This process spreads a fine sediment and sludge around the entire construction zone and into the ocean currents, which asphyxiates sea life and increases turbidity (PCA 2016, para. 857, 978–979, 982, citing Ferse Report). Entire coral colonies are destroyed, wildlife is buried and asphyxiated, and entire fish colonies, which are an intrinsically regional resource, are decimated (PCA 2016, para. 979, citing Ferse Report; McManus 2017, p. 18). Due to the size and scale of the construction area, there is also an inability of species to escape harm.

Since the island is not a natural occurrence, continuous maintenance and "re-nourishment" is also needed to prevent natural erosion; therefore, there will be continuous disturbance of any ecosystem that attempts to regenerate (PCA 2016, para. 982, citing Ferse Report). This, alongside the daily life of the new, human inhabitants of the island, who have food, transportation, and sanitations needs, will also add to the thermal distress to any species in the area, as well as increase air, land, marine, noise, and light pollution (PCA 2016, para. 977, citing Ferse Report).

Largely, it appears that the PRC either denies doing harm, in the face of contrary evidence, or seems to rely on bad science (PCA 2016, para. 982):

> As to the general claim that the construction activity "does not damage the environment on the reefs," the Ferse Report stated it "is contradicted by the facts." While the Ferse Report noted that the Chinese statements contained "accurate descriptions of the environmental conditions at the reefs," the Chinese assessments of the nature and extent of impacts from construction were "largely in disagreement with the available information."

For example, they claim that they are following green construction and building "eco-friendly reefs" (PCA 2016, para. 922). Yet this is refuted by satellite imagery, scientific reports, and witness reports (PCA 2016, para. 89(b), citing Lee 2016). They also assert a hands-off technique for rehabilitation: "China takes the approach of "natural simulation... sea storms blowing away and moving biological scraps which gradually evolve into oasis on the sea" and "the natural functions of the air, the rain, and the sun, paving it with some quick man-made material, the land reclamation area will produce the ecological effects... to a green coral reef ecological environment." (PCA 2016, para. 920, 922). Indeed, they maintain that the reefs will be "notably enhanced." (PCA 2016, para. 920).

The Tribunal stated, "[t]he conclusions of the Tribunal-appointed independent experts are unequivocal, [China has] 'impacted reefs on a scale unprecedented in the region'." (PCA 2016, para. 978, citing Ferse Report).

In addition to the harm caused by the construction of the islands is the consideration that PRC fishing practices will also expand. On multiple occasions, patrolling vessels of the Philippines have caught the PRC harvesting hawksbill sea turtles and giant clams, endangered species under international law (PCA 2016, para. 952, 956–957, 960, 964, citing CITES Appendix II). And so, the PRC has now begun to escort their fishing vessels with military vessels (PCA 2016, para. 950–951, 964). In 2012, when the Philippine Coast Guard saw PRC fishing vessels with stacks of endangered giant clams, they reported "we cannot apprehend the poachers because they are being protected by two Chinese Marine Surveillance ships." (PCA 2016, para. 839).

The PRC has responded in the past that any illegal catching will be addressed by national laws and that they strictly adhere to their obligations under UNCLOS, CBD, and CITES (PCA 2016, para. 913, 963–964). They claim that they are "always against illegal fishing." (PCA 2016, para. 913(c), citing Memorandum from the Embassy of the Republic of the Philippines in Beijing to the Secretary of Foreign Affairs, Republic of the Philippines, No. ZPE-09-2001-S (17 March 2001) (Annex 47)). At the time of the decision, however, the Tribunal could find no evidence of any domestic enforcement regarding the poaching of endangered species: "...China, despite its rules on the protection of giant clams, and on the preservation of the coral reef environment generally, was fully aware of the practice and has actively tolerated it... to exploit the living resources of the reefs..." (PCA 2016, para. 965).

3.3 Geopolitical Impacts

The PRC should build common ground through diplomacy in the region, and not artificial ground through dredging in the South China Sea (USSH 2015). The rampant environmental destruction, the apparent lack of comity and respect for neighbors in the region, the violations of international law: each alone seem like a reason that would increase tensions in the South China Sea. What makes Fiery Cross Reef even more outstanding, however, is that the PRC built an airstrip there. And a military base. And anti-aircraft missiles (Bodeen 2016). Through the creation of one small, Yongshu Island, it brought the military might of the PRC to the doorsteps of its neighbors (Bodeen 2016). The militarization of the South China Sea was discussed above in the context of physical security, so this section will discuss more the implications to foreign relations.

The PRC continually criticizes international attention to this region, and instead argues for bilateral negotiations with its neighbors (Chan 2016, p. 34). The PRC Foreign Ministry has made clear that "China always stands that, with regard to the relevant disputes between China and the Philippines in the South China Sea, a true solution can only be sought through bilateral negotiation and consultation." (PCA, para. 97). Some negotiations experts see this as a strategy to focus on the weakest party, neutralize them, and move on (Chan 2016, p. 34).

The people of the Philippines have been strongly protesting the PRC's actions for decades, and have robustly supported nature conservation in the region (Claparols 2016). By initiating proceedings with the Tribunal, the Philippines was one of the first nations to draw attention to the PRC's environmental destruction: "...a great and irreversible ecological harm" (PCA 2016, para. 680). And Philippines Supreme Court Justice Antonio Carpio has called the PRC's actions, "a grand theft of the global commons." (Lim 21 Aug 2015a). Justice Carpio, along with many non-governmental organizations (NGOs), have also been the largest advocates for creating a Marine Peace Park in the contested waters, with cooperative, regional management (Claparols 2016). *Philippines v. China* was also seen as a major victory for international law, for the Philippines, and for regional security.

However, recently, the government of the Philippines, a long-time ally of the United States, seems to be making significant changes in its domestic and foreign policy relating to the South China Sea; in effect, moving away from a partnership with the U.S. and into a closer partnership with the PRC (Perlez 2016). The news and rhetoric changes daily and the legal ramifications of Philippines President Duterte's statements are unclear, but fishermen from the Philippines were recently allowed to return to fish in Scarborough Shoal (Hunt and Quiano 2016). This is an area within the Philippines EEZ that the PRC has occupied: "Scarborough Shoal is part of Chinese territory and . . . Chinese fishermen have been fishing in the area since ancient times" (PCA 913(b), citing Memorandum from the Assistant Secretary for Asian and Pacific Affairs, Department of Foreign Affairs, Republic of the Philippines, to the Secretary of Foreign Affairs, Republic of the Philippines (14 February 2001) (Annex 45)).

The United States is perhaps the most important indirect party to China's disputes in the South China Sea (Chan 2016, p. 106). As discussed above in context with human security, and again under UNCLOS, U.S. efforts in the region are intended to be acts of deterrence to prevent open hostilities, as well as to ensure Freedom of Navigation (FON). In nearly every U.S. Congressional hearing on the South China Sea, leaders and experts go on record to emphasize the importance of peace, diplomacy, and de-militarization in the region (USSH 2015). However, it is crucial that the United States be aware of the potential moral hazard in their actions to deter militarization, as by their very presence, they may be producing the behavior that they seek to deter (Chan 2016, p. 32).

In regards to regional alliances, the PRC is not a party to the Association of South East Asian Nations (ASEAN), but it does sometimes work with the regional organization (ASEAN 2016). The group is split on the claims in the South China Sea, but together with the PRC, they did adopt a Declaration on the Conduct (DOC) of the Parties in the South China Sea in 2002 (ASEAN 2002). The DOC called on parties to recognize their binding commitments and the rule of law, to use the appropriate dispute resolution mechanisms, and to not take unilateral action to assert claims. In 2012, on the 10-year anniversary, ASEAN came together again with China and released a Joint Statement, "Enhancing Peace, Friendship, and Cooperation." (ASEAN 2016). Some ASEAN parties are now trying to develop a new, binding document (McManus 2017, p. 27). In August 2015, ASEAN Secretary General Le Luong Minh released a statement, "[w]e are calling for the termination of such activities [by China], which are of concern to us, and eroding trust and confidence among the parties, and complicating the very process of negotiating. In the face of the situation, it is even more urgent for ASEAN and China to early conclude the [new] DOC." (Lim 4 Aug 2015b).

4 A Way Forward

The PRC is a nation among nations, and several nations have claims in this region. Until determined by an appropriate dispute resolution body, or through bilateral negotiations, no claim should have priority over another. The importance of this region to the marine environment and to human security is of local, regional, and global importance, and so any actions in this area will and should draw local, regional, and international attention or condemnation. This is how societies work together when living on one planet, and under the auspices of international law.

Several nations, scientists, and NGOs have been working together to offer peaceful ways forward, based on regional cooperation and management, that focuses on shelving disputes and protecting the marine environment for current and future generations (McManus 2017, pp. 25–30). For example, Taiwan and Japan have had success with cooperative dispute resolution through the South China Sea Peace Initiative (Republic of China (Taiwan) Ministry of Foreign Affairs; McManus 2017, citing Taiwan Ministry of Foreign Affairs; Tsai 2015).

It could be particularly beneficial if the PRC saw the region as an opportunity to be a leader in regional conservation. Together, with the other nations in the region, they could combine resources and expertise to prioritize the rehabilitation of the marine environment around Fiery Cross Reef and the other occupied reefs. Also, as other nations are starting to build artificial islands, they could review their current practices and better plan for the future; such as, what is the long-term effect of these structures on the surrounding ecosystems? Can these structures survive rising sea levels due to climate change? They could do this through a regional organization like ASEAN, or create their own body specific to this task. A new and binding Declaration of Conduct could also show good faith intent, and serve as the foundation for an ethical, just, and fair regional management program for the waters and resources of the South China Sea.

There is an inextricable link between the environment and human security, and environmental rights are human rights. In this region, rich in biodiversity, cultural diversity and history, a regional approach based in mutual respect, and mutual needs, is the best path forward for the future and flourishing of each state, of the region, and of the planet. One should not destroy the foundations for security, in one's quest for security.

References

Associated Foreign Press (AFP) (16 Dec 2016) China seizes US naval drone in South China Sea. Al Jazeera. Available at: http://www.aljazeera.com/news/2016/12/china-seizes-naval-drone-south-china-sea-161216172602149.html. Accessed 17 Dec 2016

Bale R (29 Aug 2016) One of the World's biggest fisheries is on the verge of collapse. National Geographic. Available at: http://news.nationalgeographic.com/2016/08/wildlife-south-china-sea-overfishing-threatens-collapse/. Accessed 24 Oct 2016

Bartley A (17 Nov 2016) The secret driver of the South China Sea disputes: China's hunger for fish. The Diplomat. Available at: http://thediplomat.com/2016/11/the-secret-driver-of-the-south-china-sea-disputes-chinas-hunger-for-fish/. Accessed 17 Dec 2016

Batongbacal J (7 May 2015) Environmental aggression in the South China Sea. Asia Maritime Transparency Initiative. Available at: https://amti.csis.org/environmental-aggression-in-the-south-china-sea/. Accessed 24 Oct 2016

Bodeen C (15 Dec 2016) Report: Beijing adds weapons to South China Sea islands. Yahoo News. Available at: https://www.yahoo.com/news/us-report-says-china-adds-weapons-south-china-042523142.html. Accessed 17 Dec 2016

Bohane B (8 Dec 2016) South China Sea: Vietnam prepares for dangerous days ahead as the country's fisheries clash with chinese authorities. ABC News. Available at: http://www.abc.net.au/news/2016-12-09/south-china-sea-vietnam-prepares-for-dangerous-days-ahead/8101192. Accessed: 17 Dec 2016

Callahan M (22 Oct 2016) US destroyer sails in South China Sea to make a point. CNN. Available at: http://www.cnn.com/2016/10/21/politics/us-guided-missile-destroyer-south-china-sea/index.html. Accessed 17 Dec 2016

Chan S (2016) China's troubled waters: maritime disputes in theoretical perspective. Cambridge University Press, Cambridge

Central Intelligence Agency (CIA) (2016) World Fact Book: Spratly Islands. Available at: https://www.cia.gov/library/publications/the-world-factbook/geos/pg.html. Accessed: 17 Dec 2016

Claparols A (27 Oct 2016) Sad times for conservation. Philippine Daily Inquirer. Available at: http://opinion.inquirer.net/97469/sad-times-for-conservation. Accessed 24 Oct 2016

Cochrane J (20 June 2016) Indonesia confirms seizing fishing boat in South China Sea, defying Beijing. NY Times. Available at: http://www.nytimes.com/2016/06/21/world/asia/indonesia-south-china-sea-fishing.html?_r=0

Denyer S (12 April 2016) How China's fishermen are fighting a covert war in the South China Sea. Available at: https://www.washingtonpost.com/world/asia_pacific/fishing-fleet-puts-china-on-collision-course-with-neighbors-in-south-china-sea/2016/04/12/8a6a9e3c-fff3-11e5-8bb1-f124a43f84dc_story.html?utm_term=.ab3a76991022. Accessed 17 Dec 2016

Glaser B (27 Oct 2015) The US Asserts Freedom of Navigation in the South China Sea. Center for Strategic and International Studies. Available at: https://www.csis.org/analysis/us-asserts-freedom-navigation-south-china-sea. Accessed 18 Dec 2016

Henkin L (1979) How nations behave. Columbia University Press, New York

Henkin L (1987) Restatement (third) of the foreign relations law of the United States § 102(2). American Law Institute, Philadelphia

Hunt K, Quiano K (31 Oct 2016) China allows Philippines fishermen access to disputed shoal in South China Sea. Available at: http://www.cnn.com/2016/10/31/asia/philippines-china-scarborough-shoal-fishermen/index.html. Accessed 18 Dec 2016

Kintzele K (2011) Keeping Nature Alive: from moral motivations to legal implications. Globalisation and Ecological Integrity. In: Westra L, Bosselmann K, Soskolne C (eds) Science and international law. Cambridge Scholars Publishing, Newcastle-upon-Thyne

Lee VR (15 Jan 2016) Satellite Imagery Shows Ecocide in the South China Sea. The Diplomat. Available at thediplomat.com/2016/01/satellite-images-show-ecocide-in-the-south-china-sea/. Accessed 24 Oct 2016

Lim K (27 May 2016) G7 Statement Ignores Beijing's Request to Stay Mum on South China Sea. Available at: http://www.inquisitr.com/3139267/g7-statement-ignores-beijings-request-to-stay-mum-on-south-china-sea/. Accessed 24 Oct 2016

Lim K (4 Aug 2015a) Let It Go: ASEAN countries Entreat China regarding its takeover of the South China Sea. Available at: http://www.inquisitr.com/2309103/let-it-go-asean-countries-entreat-china-regarding-its-takeover-of-the-south-china-sea/. Accessed 24 Oct 2016

Lim K (21 Aug 2015b) The greening of the Philippines. Justice Carpio: Fish Sanctuary Could Ease Philippine-China Conflict. Available at: http://www.philippinesgreening.com/justice-carpio-fish-sanctuary-could-ease-philippine-china-conflict/. Accessed 24 Oct 2016

Malig J (6 June 2013) UNCLOS does not support China's claims, US naval expert says. ABS CBN News. Available at: http://news.abs-cbn.com/focus/06/06/13/unclos-does-not-support-chinas-claims-us-naval-expert-says. Accessed 18 Dec 2016

Manyin M, Daggett S, Dolven B, Lawrence SV, Martin MF, O'Rourke R, Vaughn B (28 Mar 2012) Pivot to the Pacific? The Obama Administration's "rebalancing" toward Asia. Congressional Research Services. Available at: http://fas.org/sgp/crs/natsec/R42448.pdf. Accessed 24 Oct 2016

McManus J (2017) Offshore coral reef damage, overfishing and paths to peace in the South China Sea. International Journal of Marine and Coastal Law: Special Issue on the South China Sea

Mollman S (7 June 2016) One of Beijing's controversial manmade islands in the South China Sea now has a farm. Available at: http://qz.com/700977/one-of-beijings-controversial-manmade-islands-in-the-south-china-sea-now-has-a-farm/. Accessed 17 Dec. 2016

Permanent Court of Arbitration (PCA-CPA) (2016) About us Available at: https://pca-cpa.org/en/about/. Accessed 17 Dec 2016

Permanent Court of Arbitration (PCA) (12 July 2016) Philippines v. China. Available at: https://pca-cpa.org/wp-content/uploads/sites/175/2016/07/PH-CN-20160712-Award.pdf. Accessed 24 Oct 2016

Perlez J (20 Oct 2016) Rodrigo Duterte and Xi Jinping agree to reopen South China Sea talks. New York Times. Available at: http://www.nytimes.com/2016/10/21/world/asia/rodrigo-duterte-philippines-china-xi-jinping.html?_r=0. Accessed 24 Oct 2016

Phillips T, Holmes O, Bowcott O (12 July 2016) Beijing rejects tribunal's ruling in South China Sea case. The Guardian. https://www.theguardian.com/world/2016/jul/12/philippines-wins-south-china-sea-case-against-china. Accessed 18 Dec 2016
Republic of China (Taiwan) Ministry of Foreign Affairs. South China Sea Peace Initiative. Available at: http://www.mofa.gov.tw/en/theme.aspx?n=E5A0D5E2432C234D&s=83376F561B7165E6&sms=BCDE19B435833080. Accessed 18 Dec 16
Rosenberg D (2016) The South China Sea: Territorial Claims Maps. Available at: http://www.southchinasea.org/maps/territorial-claims-maps/. Accessed 18 Dec 2016
Sharman J (11 Dec 2016) China flies nuclear bomber over South China Sea as a 'message' to Donald Trump. Available at: http://www.independent.co.uk/news/world/asia/china-bomber-flight-send-message-donald-trump-taiwan-a7468021.html. Accessed 18 Dec 2016
Shicun W (6 Mar 2014) Is the nine-dash line in the South China Sea legal? China-US Focus. Available at: http://www.chinausfocus.com/peace-security/is-the-nine-dash-line-in-the-south-china-sea-legal/. Accessed 18 Dec 2016
The Association of Southeast Asian Nations (ASEAN) (2002) Declaration on the Conduct of the Parties of the South China Sea. Available at: http://asean.org/?static_post=declaration-on-the-conduct-of-parties-in-the-south-china-sea-2. Accessed 24 Oct 2016
The Association of Southeast Asian Nations (ASEAN) (2016) External Relations: China. Available at: http://asean.org/asean/external-relations/china/. Accessed 18 Dec 2016
Tsai G (28 Aug 2015) Taiwan and the South China Sea peace initiative. The Diplomat. http://thediplomat.com/2015/08/taiwan-and-its-south-china-sea-peace-initiative/. Accessed 18 Dec 2016
Treves T (2008) UNCLOS introductory note. UN Audiovisual Library of International Law. Available at: http://legal.un.org/avl/ha/uncls/uncls.html. Accessed 18 Dec 2016
Trowbridge L (2016) South China Sea, between the Philippines, borneo, Vietnam, and China. World Wildlife Fund. Available at: http://www.worldwildlife.org/ecoregions/im0148. Accessed 17 Dec 2016
United Nations Conference on Environment and Development (UNCED) (2016) Available at: http://www.un.org/geninfo/bp/enviro.html/. Accessed 17 Dec 2016
United Nations Convention on the Law of the Sea (UNCLOS) (1982) Available at: http://www.un.org/Depts/los/convention_agreements/texts/unclos/unclos_e.pdf. Accessed 24 Oct 2016
United Nations Development Program (UNDP) (1994) Human development report. Oxford University Press. Available at: http://hdr.undp.org/sites/default/files/reports/255/hdr_1994_en_complete_nostats.pdf. Accessed 24 Oct 2016
United Nations General Assembly (UNGA) (10 Sept 2012) Resolution adopted by the General Assembly on 10 September 2012, 66/290. Follow-up to paragraph 143 on human security of the 2005 World Summit Outcome. Available at: http://www.un.org/ga/search/view_doc.asp?symbol=A/RES/66/290. Accessed 17 Dec 2016
United Nations Trust Fund for Human Security (2016) Available at: http://www.un.org/humansecurity/. Accessed 17 Dec 2016
U.S. Department of Defense (USDOD) (5 Jan 2012) Defense Strategic Guidance Briefing from the Pentagon. Available at: http://archive.defense.gov/Transcripts/Transcript.aspx?TranscriptID=4953. Accessed 17 Dec 2016
U.S. Department of Defense (USDOD) (2015) The Asia-Pacific Maritime Security Strategy: Achieving U.S. National Security Objectives in a Changing Environment. Available at: http://www.defense.gov/Portals/1/Documents/pubs/NDAA%20A-P_Maritime_SecuritY_Strategy-08142015-1300-FINALFORMAT.PDF. Accessed 24 Oct 2016
U.S. Department of State (USDOS) (2016) Maritime Security and Navigation. Available at: http://www.state.gov/e/oes/ocns/opa/maritimesecurity/. Accessed 17 Dec 2016

U.S. Senate Hearing (USSH) (13 May 2015) Safeguarding American Interests in the East and South China Seas. Hearing before the Committee on Foreign Relations. 114-75: 114th Congress, 1st Session, Washington DC

Westra L (1994) An environmental proposal for ethics: the principle of integrity. Rowman and Littlefield Publishers, London

Wee S-L (2014) Chinese ships patrol area contested by Malaysia. Reuters. Available at: http://uk.reuters.com/article/uk-china-malaysia-idUKBREA0P06X20140126. Accessed 18 Dec 2016

Weeramantry C (1997) Gabčíkovo–Nagymaros project (Hungary/Slovakia), separate opinion of vice president Weeramantry. International Court of Justice. Available at: http://www.icj-cij.org/docket/files/92/7383.pdf. Accessed 18 Dec 2016

Part III
Responsibility for Human Rights Breaches and Climate Change

The Enormity of the Damage Done by the Climate Change Disinformation Campaign as the World Struggles to Implement the Paris Agreement

Donald A. Brown

1 Introduction

This chapter explains the enormous damage to people around the world and ecological systems on which hundreds of millions depend caused by a climate change disinformation campaign which has successfully blocked government action to reduce the threat of climate change for almost 50 years in the United States and several other developed countries. This damage is most notable in light of the staggering challenge to the world of limiting future warming to non-catastrophic levels as the international community strives to implement the Paris Agreement (UNFCCC 2015) which was concluded at the twenty-first Conference of the Parties (COP-21) under the United Nations Framework Convention on Climate Change (UNFCCC 1992) in December, 2015.

The chapter begins with a description of the climate change disinformation campaign, its tactics, and its successes in blocking meaningful national responses to climate change since the 1980s in the United States and in several other developed countries. The chapter then describes the enormity of the challenge facing the world to prevent catastrophic climate change as the international community strives to implement the 2015 Paris Agreement to prevent catastrophic warming.

D.A. Brown (✉)
Widener University Commonwealth Law School, Harrisburg, PA, USA
e-mail: dabrown57@gmail.com

L. Westra et al. (eds.), *The Role of Integrity in the Governance of the Commons*,
DOI 10.1007/978-3-319-54392-5_8

2 The Climate Change Disinformation Campaign

The sociological literature on the disinformation campaign describes this phenomenon as a counter-movement. (See, for example, McCright and Dunlap 2000, p. 559) A counter-movement is a social movement that has formed in reaction to another movement (McCright and Dunlap 2000, p. 504). The climate change disinformation campaign can be understood to be a continuation of the counter-movements that arose among US political conservatives in reaction to the environmental, civil rights, women's rights, and anti-war movements that arose in the 1960s in the United States. And so, the climate change disinformation campaign's methods and processes can be understood to be an extension of strategies that had already been developed among some, although not all, conservatives to counter the environmental movement that had developed in the late 1960s and 1970s around other environmental issues such as air and water pollution, safe disposal of waste and toxic substances, and protection of wetlands and endangered species.

Yet the emergence of global warming as an issue in the 1980s with its potential for large-scale social change needed to ameliorate its threat was seen as more threatening to conservatives in regard to industry, prosperity, life-style, and the entire American-way of life, than were traditional pollution problems (McCright and Dunlap 2000, p. 503) In other words, climate change directly threatened the central values of the US conservative movement even more than other environmental problems (McCright and Dunlap 2000, p. 505). Climate change policies also greatly threatened the profits of the most powerful industry in the world, the fossil fuel industry. As a result climate change has become the key environmental focus of the US conservative movement and corporate opponents of environmental regulation. In subsequent years the disinformation campaign would be taken up in other countries including the United Kingdom and Australia.

The climate change disinformation movement can be understood to be comprised of many organizations and participants including conservative think tanks, front groups, Astroturf groups, conservative media, public relations firms, and individuals. This disinformation campaign frequently has used a number of tactics to convince people and politicians that the science supporting climate change policies is flawed. The central claims of the climate change disinformation movement have been:

- There is no warming.
- Its not caused by humans.
- Reducing greenhouse gas emissions will cause more harm than good.
  (McCright and Dunlap 2010, p. 111)

To support these basic claims, the climate denial machine frequently has made claims that mainstream climate scientists are corrupt or liars, descriptions of adverse climate change impacts are made by "alarmists" scientific journals that publish climate related research are biased against skeptics, and mainstream climate science is "junk" science. The climate change disinformation machine also has made frequent ad hominem attacks on those who produce climate change science and sometimes has cyber-bullied both climate scientists and journalists. (For a discussion of cyber-bullying of scientists and journalists by some participants in the climate change disinformation campaign see, Brown 2012b).

The climate change disinformation campaign began in the 1980s when some of the same scientists and organizations that fought government regulation of tobacco began to apply the tactics perfected in their war on the regulation of tobacco to climate change. (Oreskes and Conway 2010, pp. 169–215).

According to Pooley, the climate disinformation campaign began "spinning around 1988 in response to the increasingly outspoken scientific community..." (Pooley 2010, p. 39). Since then for almost 25 years this campaign has been waged to undermine support for regulation of GHGs.

To say that the campaign has been "waged" is not to claim that it has been a tightly organized, completely coordinated effort by a few groups or individuals or that all participants have the same motives. In fact, different participants may have radically different motives including the fact that some may be sincere, some appear to be motivated by protecting free markets without government intervention, and many appear to believe that no restriction on fossil fuel use can be justified without very high levels of proof of harms because of the economic power of the fossil fuel industry. Yet, these different participants since the 1990s for the most part have acted in a well-coordinated campaign among contrarian scientists, free-market think tanks, and industry to create a fog of doubt around climate change (Begley 2007). They have accomplished this through the production of advertisements, op-eds, lobbying, books, media attention, and quotations from skeptical scientists often associated with conservative think tanks. They have argued first that the world is not warming, measurements that indicate otherwise are flawed, any warming is natural, that is not caused by human activities, and if warming does occur it will be miniscule and harmless (Begley 2007).

Different groups created this counter-movement often acting independently of each other, yet connected through the internet to create a denial machine that has effectively responded to any public pronouncement by scientists or journalists that have asserted that human-induced climate change is a serious problem (Begley 2007). Conservative activists wrote hundreds of documents (including policy briefs, books, press releases, and op-eds), held numerous policy forums and press conferences, appeared regularly on television and radio programs, and testified at congressional hearings on global warming (Dunlap and McCright 2008).

As a result of the internet communication between participants in this campaign, charges by one of the participants have been quickly transmitted to others creating an echo chamber of counter-claims made in opposition to the mainstream scientific view of climate change.

A recent book edited by sociologists Riley Dunlap and Robert Bruelle on the sociology of climate change describes the rise of a climate change disinformation campaign in the United States and its export to several other developed countries (Dunlap and Brulle 2015). The climate change countermovement arose when those corporations and organizations who were threatened by calls for governments to take action to reduce the threat of climate change organized themselves to protect their economic and political interests that would be threatened by regulation of fossil fuels.

Although the funding for the disinformation campaign came initially mostly from numerous fossil fuel corporations and free market fundamentalist foundations, between 2005 and 2008, a single source, the Koch brothers, provided almost $25 million into dozens of different organizations fighting climate change policies (Mayer 2016, p. 204).

According to sociologist Robert Brulle, 140 conservative foundations funded the disinformation campaign by distributing $558 million in the form of 5299 grants to 91 different nonprofit organizations engaged in opposition to climate change policies (Brulle 2014, p. 694).

Although responsible scientific skepticism about climate change should be encouraged, the climate disinformation campaign can be understood as a movement of corporations, organizations, and individuals that have systematically attacked mainstream climate change science using tactics that are radically inconsistent with responsible scientific skepticism. The tactics of the climate change disinformation campaign have been documented to include:

- Lying or reckless disregard for the truth.
- Focusing on unknowns while ignoring the knowns,
- Specious claims of "bad science" about any conclusion that is not fully proven,
- Creation of front groups or "Astroturf" groups which hide the real parties in interest,
- Manufacturing bogus climate science by inviting skeptics to present papers on climate science that have not been subject to "peer-review" and then widely publicizing these papers as if they were sound scientific conclusions,
- Think Tank campaigns that publicize non-peer reviewed skeptical climate science.
- Misleading Public Relations campaigns that target uneducated citizens,
- Cyber Bullying scientists and journalists

  (Brown, 2011)

These tactics do no constitute responsible scientific skepticism but morally reprehensible disinformation (For a discussion of this see Brown 2010).

The emergence of global warming as an issue in the 1980s with its potential for large-scale social change needed to ameliorate its threat was seen as more threatening to conservatives in regard to industry, prosperity, life-style, and the entire American-way of life, than were traditional pollution problems (McCright and Dunlap 2000 p. 503). In other words, climate change directly threatened the central values of the US conservative movement even more than other environmental problems (McCright and Dunlap 2000, p. 505). As a result climate change has become the key environmental focus of the US conservative movement.

In subsequent years the climate change disinformation campaign would be taken up in other countries including the United Kingdom and Australia.

Different groups created this counter-movement often acting independently of each other, yet connected through the internet to create a denial machine that has effectively responded to any public pronouncement by scientists or journalists that have asserted that human-induced climate change is a serious problem (Begley 2007). Conservative activists wrote hundreds of documents (including policy briefs, books, press releases, and op-eds), held numerous policy forums and press conferences, appeared regularly on television and radio programs, and testified at congressional hearings on global warming (Dunlap and McCright 2008).

As a result of the internet communication between participants in this campaign, charges by one of the participants have been quickly transmitted to others creating an echo chamber of counter-claims made in opposition to the mainstream scientific view of climate change.

The disinformation campaign's most important participants have been conservative think tanks according to the sociological literature (Jacques et al. 2008). These think tanks developed the ideas, communications and media strategies, literature and press releases that have been widely deployed in rhetorical strategies to defend conservative interests by creating doubt about mainstream climate change scientific claims.

Initially most of the funding for this disinformation campaign came from fossil fuel interests and corporations whose products produce high levels of GHGs. On October 21, 2010, John Broder of the New York Times reported that:

> the fossil fuel industries have for decades waged a concerted campaign to raise doubts about the science of global warming and to undermine policies devised to address it (Broder 2010).

According to Broder, the fossil fuel industry has:

> created and lavishly financed institutes to produce anti-global-warming studies, paid for rallies and Web sites to question the science, and generated scores of economic analyses that purport to show that policies to reduce emissions of climate-altering gases will have a devastating effect on jobs and the overall economy (Broder 2010).

Not surprisingly, the fossil fuel industry funded many of the initial efforts to prevent adoption of climate change policies. Both individual corporations such as ExxonMobil and Peabody Coal, as well as industry associations such as American Petroleum Institute, Western Fuels Associations, and Edison Electric Institute provided funding for individual contrarian scientists, conservative think tanks active in climate change denial, and a host of front groups that we will discuss below (Dunlap and McCright 2011, p. 148),

Although the initial funding in the campaign came mostly from a number of fossil fuel corporations, McCright and Dunlap argue that more recently conservative, free-market, and anti-regulatory foundations have been the main forces fueling the denial machine. (Dunlap and McCright 2011, p. 144).

According to sociologists Dunlap and McCright the glue that holds the elements of the climate disinformation campaign together is a shared hatred for government regulation of private industry (Dunlap and McCright 2011, p. 144) And so, a staunch commitment to free markets and a disdain for government regulation are the ideas that most unite the climate denial community (Dunlap and McCright 2011, p. 144).

Some fossil fuel corporations more than others have been key funders among corporate funders of the disinformation campaign. A recent study published in the Proceedings of the National Academies of Science shows that the climate denial echo chamber organizations were funded primarily by ExxonMobil and Koch family foundations and produced misinformation that effectively polluted mainstream media coverage of climate science and polarized the climate policy debate (Farrell 2015). This study's analysis of 20 years' worth of communication data between participants in the climate change countermovement by Yale University researcher Dr. Justin Farrell shows beyond a doubt that ExxonMobil and the Kochs have been key actors who funded the creation of climate disinformation think tanks and ensured the prolific spread of their doubt products throughout our mainstream media and public discourse about climate change.

"The contrarian efforts have been so effective for that they have made it difficult for ordinary Americans to even know who to trust," Dr. Farrell told the Washington Post. "This countermovement produced messages aimed, at the very least, at creating ideological polarization through politicized tactics, and at the very most, at overtly refuting current scientific consensus with scientific findings of their own" (Warrick 2015).

Two main conclusions of this study are noteworthy because they are strong evidence for the role that fossil fuel interests played in creating a false cultural understanding of climate change science in the United States. The conclusions are:

> First, that organizations with corporate funding were more likely to have written and disseminated texts meant to polarize the climate change issue.
>
> Second, and more importantly, that corporate funding influences the actual thematic content of these polarization efforts, and the discursive prevalence of that thematic content over time (Farrell 2015, Abstract)

Thus this study, which was published by the very prestigious US Academy of Sciences, provides further evidence that the public deception orchestrated by the fossil fuel industry actually influenced the cultural understanding of climate change in the United States.

Those engaged in this disinformation campaign can be distinguished from responsible climate skeptics because the climate change denial campaign is a collective social movement run by professional advocacy working to discredit climate change" (Hoffman 2011, p. 5) As such, this movement is not engaged in reasonable scientific skepticism but advocacy that stresses scientific uncertainty. In fact, McCright and Dunlap summarize the disinformation machine as having been engaged in misrepresenting, manipulating, and suppressing climate change research results. (McCright and Dunlap 2010, p. 111).

Although almost all of the disinformation campaign led opposition to climate change policies has been on the basis of inadequate scientific grounding for action, scientific arguments are usually coupled with economic arguments such as claims that climate change policies will destroy jobs, hurt specific industries, lower GDP, or are not justified by cost-benefit analysis.

The original organizations that sought to undermine public support for climate policies by exaggerating scientific uncertainty have expanded to include ideological think tanks, front groups, Astroturf groups (i.e., groups organized by industry that pretend to be a legitimate grassroots organization), and PR firm led campaigns (Oreskes and Conway 2010, pp. 169–215).

Without doubt, the climate change disinformation machine has campaigned to emphasize the uncertainties associated with climate change with the intention of giving the impression that lack of certainty undermines scientific conclusions. A team assembled by the American Petroleum Institute (API) created a "Global Climate Science Action Plan" that had as its purpose to convince the public that climate science is awash in uncertainty (Hoggan 2009, p. 42) An API document describes the mission they were supporting as the following:

Victory will be achieved when:

- Average citizens understand uncertainties.
- Media understands uncertainties.
- Media coverage reflects balance on climate science and recognition of the validity of viewpoints that challenge current conventional wisdom.
- Industry senior leadership understands uncertainties in climate science, making them stronger ambassadors to those who shape climate policy.
- Those promoting the Kyoto treaty on the basis of extant science appear to be out of touch with reality.

(Hoggan 2009 p. 43)

This API document demonstrates that some participants in the climate change disinformation machine were more interested in framing climate change as uncertain than they were in educating citizens about the scientific basis for concern. This tactic must be understood to be not only greatly misleading given what is not in contention about climate change but also an attempt to manipulate people and governments to protect the economic interests of the members of the American Petroleum Institute.

Efforts to undermine climate change science which were already sporadically underway in the mid-1980s yet began to get better organized in the United States shortly after James Hansen testified in the US Senate in 1988 that climate change was already visible, testimony which put climate change squarely on the US public agenda (Dunlap and McCright 2015, p. 300). Organized climate change denial continued to grow in the next two decades and reached an unprecedented level in 2009 when the newly elected Obama administration and the Democratically controlled US Congress increased the likelihood of US action to reduce greenhouse gas (GHG) emissions with the result that no climate change legislation was enacted. These efforts have continued relatively unabated since then (Dunlap and McCright 2015, p. 300). Further, climate change denial has become a virtual "litmus test" for Republican politicians, strongly enforced by elements of the conservative movement (Dunlap and McCright 2015, p. 300).

Leading fossil fuel corporations (most notably ExxonMobil and Peabody Coal, industry associations, including the American Petroleum Institute and Western Fuels, initially led efforts to deny climate change (Dunlap and McCright 2015, p 310). These fossil fuel corporations and industry organizations were joined by many other corporations and business associations to fund climate science deniers and conservative Think Tanks and various groups promoting climate change science denial (Dunlap and McCright 2015, p. 310).

Political conservatives in the United States learned from the Reagan administration's experience that it was unwise to attack environmental protection directly, given that Americans were generally supportive of environmental protection (Dunlap and McCright 2015, p. 306). As a result, the conservatives and their industry allies learned to prevent the implementation of government policies that might threaten their political and economic interests by undermining the scientific foundations of environmental policy proposals (Dunlap and McCright 2015, p. 306). As result conservatives seized upon the strategy of "manufacturing uncertainty" that had been previously effectively employed for several decades by corporations and entire industries, most notably the tobacco industry, in efforts to protect their products from regulations and lawsuits by questioning the scientific adequacy of claims that their products were hazardous (Dunlap and McCright 2015, p. 306). Thus conservatives began labeling science supporting the need to regulate industry to protect the environment as "junk science." This strategy became the favored tactic employed by conservatives and their industry allies when government showed interest in expanding environmental regulation and the major focus of attempts to prevent the adoption of climate change policies in the early 1990s (Dunlap and McCright 2015, p. 306).

Participants in the denial movement undermined the public's confidence in climate change science by attacking the validity of climate models, the use of paleoclimate data to establish climate trends, attacked individual climate scientists and scientific institutions, published dubious non-peer reviewed climate science reports, funded self-proclaimed climate scientists who rarely published their conclusion in peer-reviewed journals, and many other tactics that manufactured scientific uncertainty.

The denial countermovement has evolved, changed, and expanded over the past quarter-century, changes that included new key actors, supporters, and tactics while the basic strategy of manufacturing uncertainty has expanded into manufacturing public controversy about climate science up until the present (Dunlap and McCright 2015, p. 309).

The major participants in the denial countermovement have included portions of the fossil fuel industry and corporate America, conservative think tanks, a relatively small number of contrarian scientists, front groups and Astroturf organizations, conservative politicians and media, and the denial blogosphere (Dunlap and McCright 2015, p. 309).

Although the denial countermovement began in the United States it diffused internationally to countries including the United Kingdom, Canada, Australia, and

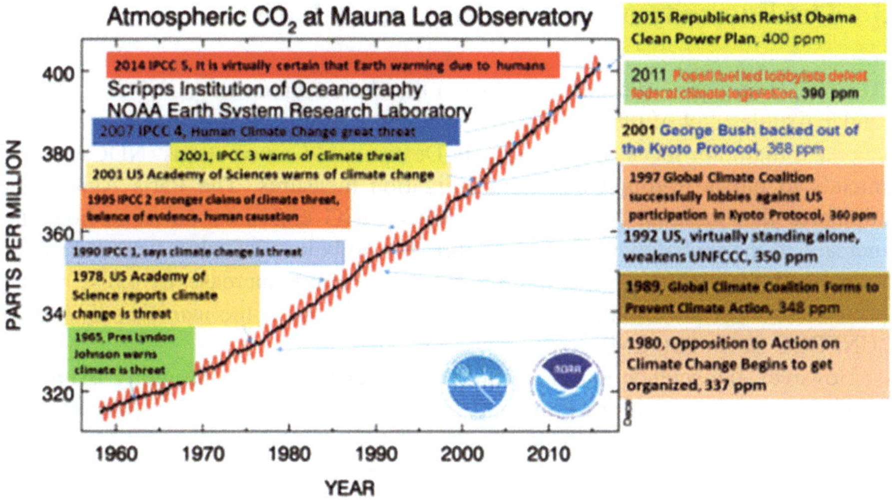

Fig. 1 Atmospheric CO_2 at Mauna Loa Observatory—Available at: Ethicsandclimate.org

recently into several European countries including France, Sweden, and the Netherlands (Dunlap and McCright 2015, p. 316).

The red saw-tooth like line in the center of Fig. 1 depicts rising atmospheric concentrations of CO_2 rising from about 320 ppm in 1965 when US President Lyndon Johnson acknowledged that climate change was a growing threat to a current atmospheric concentration of over 400 ppm CO_2. In a very abbreviated manner, Fig. 1 depicts how as prestigious scientific institutions issued reports on the danger of climate change with increasing levels of scientific certainty, the climate change disinformation blocked climate change policies. The boxes to the right of the right axis in Fig. 1 depict successful efforts of the disinformation campaign to defeat proposed climate policies at the national and international level. To the left of the line depicting rising CO_2 concentrations are depictions of some of many scientific reports issued by prestigious scientific institutions including the United States Academy of Sciences and the Intergovernmental Panel on Climate Change which all concluded with stronger levels of scientific certainty that climate change was a growing menace to the world. From this one can deduce that the climate change disinformation campaign was responsible for almost a 50-year delay in responding to the growing threat of climate change and a rise in atmospheric concentrations of CO_2 from approximately 320 ppm to current levels of slightly above 400 ppm. (For a much more rigorous analysis of the role of the climate change policy opposition in US climate policy formation see Brown (2002), Chap 2 and Brown (2012a), Chap 2 and numerous articles under the category "disinformation" on the website Ethicsandclimate.org).

In the next section, this chapter describes the enormity of the challenge facing the world caused by this 50-year delay given the staggering state of affairs facing the international community after the COP-21 in Paris in 2015.

3 The Enormity of the Challenge After Paris

At COP 21 in Paris, 195 nations adopted the Paris Agreement (UNFCCC 2015). Under Articles 2–4 of the Agreement, each Party is obliged to prepare, communicate, and maintain successive Nationally Determined Contributions (NDCs) that it intends to pursue through domestic mitigation measures. The Paris Agreement stipulates that national NDCs should reduce GHG emissions so as to hold the increase in the global average temperature to well below 2°C above pre-industrial levels and to pursue efforts to limit the temperature increase to 1.5°C above pre-industrial levels, a requirement this paper refers to as the *warming limit goal* (UNFCCC 2015, Art. 2–4).

However, the science is now clear that the international community is rapidly running out of time to limit warming to 2°C and is even more rapidly approaching a time when it will be virtually impossible to limit warming to 1.5°C (Tschakert 2015). The Paris Agreement, reaffirming national responsibilities under the 1992 United Nations Framework Convention on Climate Change (UNFCCC 1992, gations on nations to construct their emissions targets to achieve the warming limit of 1.5–2°C on the basis of "equity and common but differentiated responsibilities and respective capabilities" (UNFCCC 2015, Art. 2, para. 2).

In 2010, the international community originally agreed to limit warming to 2°C at COP-16 in Cancun Mexico (Carbon Brief 2014). At that time the international community believed that if global warming could be limited to 2°C, catastrophic warming could be avoided. However, just 5 years later in Paris, the international community agreed to limit the increase in global average temperatures to "well below 2°C and to pursue efforts to limit temperature increases to 1.5°C". The warming limit goal was lowered because recent scientific evidence has made it evident that a 2°C warming limit may not prevent dangerous climate change. In fact, some of the harsh climate impacts that were not expected until global temperatures rose 2°C were already being experienced by global temperature increases of slightly more than 1°C.

In Paris, the world's most vulnerable nations demanded that the international warming limit goal be reduced to 1.5°C. As a result, largely because of significant pressure from Small Island Developing States (SIDS) and the most vulnerable Least Developed States (LDS) who successfully argued that a warming limit of greater than 1.5°C would leave them exposed to dire climate impacts, including in some cases existential threats to their sovereignty and existence, every nation in the world agreed in Paris in 2015 to work cooperatively in an attempt to limit warming to 1.5°C. (For a description of the Paris negotiation dynamics that led to agreement on the 1.5°C warming limit, see: Obergassel et al. Obergassel (ne Sterk) et al. 2016, pp. 15–16).

Largely missing from international press coverage of the Paris agreement is the enormity of the challenge facing the world to have any hope of limiting warming to close to 1.5°C. The international press also largely ignored that the enormity of the challenge facing the world to prevent dangerous climate change had been caused by the long delay due to the success of the disinformation campaign in preventing the United States and other high emitting developed nations from adopting meaningful climate policies.

To operationalize a warming limit, it is necessary to identify a carbon budget, or the total amount of CO_2 equivalent gases (CO_2-e), that can be emitted into the atmosphere before atmospheric concentrations of GHGs exceed a level that will cause the warming limit to be exceeded. Therefore, for instance, to limit warming to 1.5°C, a global carbon budget must be calculated that will limit atmospheric CO_2-e to a concentration that is assumed will limit warming to 1.5°C. Yet the carbon budgets that remained to limit warming to between 1.5 and 2.0°C are so small that emissions reductions pathways that the international community must implement to achieve the warming limits are civilization challenging.

To illustrate this point, Fig. 2 depicts global contraction pathways consistent with carbon budgets for: (a) dotted line at far left, a 1.5°C warming limit with a 50%

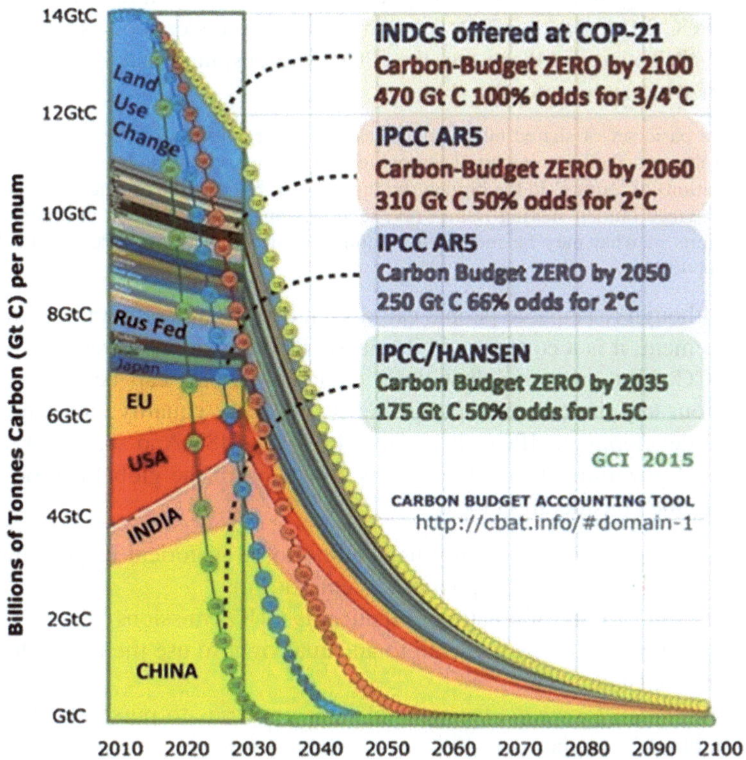

Fig. 2 INDC Emission-Control offers, 2020 to 2030 & to 2100 for 3 to 4°C Compared to Global Carbon Budgets for 1.5 and 2.0°C In relation to the 'Paris Agreement' at COP-21 (December 2016) For a clearer display of the ideas dipicted in this figure see http://www.gci.org.uk/images/INDCs_COP_21_a.pdf

probability; (b) second dotted line from the left, a 2°C limit with a 66% probability; (c) third dotted line from left, a 2°C limit with a 50% probability; and (d) dotted line at far right, pathway consistent with aggregate emissions reduction INDC pledges submitted to the UNFCCC before Paris which are believed would achieve a 3–4°C warming limit if achieved and extended into the future.

The steepness of the emissions GHG reduction curves depicted in Fig. 2 demonstrates the enormity of global GHG emissions reductions now necessary to limit warming to between 1.5 and 2.0°C.

Yet these GHG emissions reduction curves are what is required of the entire world. Developed nations will need to reduce their GHG emissions at even faster rates to have any hope of limiting global GHG emissions at the rates depicted in Fig. 2. As noted above, nations agreed to reduce their emissions to prevent dangerous climate change on the basis of equity and the principle of common but differentiated responsibilities and respective capabilities, in the light of different national circumstances (UNFCCC 2015, Art. 2. Para 2). Thus, the reduction curves required of developed countries need to be even steeper than the already steep curves in Fig. 2 to take into account equity.

In its 5th Assessment, the IPCC explained that equity was a fair burden sharing concept (IPCC 2014, Sec. 4.6.2.1). As such it should be understood as a synonym of "distributive" justice. Furthermore, the IPCC said despite a lack of unambiguous meaning of what equity means:

> there is a basic set of shared ethical premises and precedents that apply to the climate problem that can facilitate impartial reasoning that can help put bounds on the plausible interpretations of 'equity' in the burden sharing context. Even in the absence of a formal, globally agreed burden sharing framework, such principles are important in establishing expectations of what may be reasonably required of different actors (IPCC 2014, Sec. 4.6.2.1, p. 48).

Thus, although reasonable people can disagree on what equity means under the Paris Agreement, it is a concept with limits on its potential interpretation.

The IPCC went on to say that these equity principles can be understood to comprise four key dimensions: responsibility, capacity, equality, and the right to sustainable development (IPCC 2014, Sec. 4.6.2.1. p. 48). And so, step 4 requires each national government to determine its allocation of the global carbon budget after giving due consideration to:

1. *Responsibility*—which is usually understood to be historical responsibility for current elevated GHG atmospheric concentrations;
2. *Capacity*—the ability of a nation to reduce its GHG emissions.
3. *Equality*—the right of each citizen to an equal right to use the atmosphere as a sink for its GHGs;
4. *Sustainable development*—the right of poor countries to pursue economically sustainable development.

Any reasonable interpretation of what equity requires under the Paris Agreement will require the high emitting developed nations to reduce their GHG emissions at even steeper rates than the enormously challenging emissions reduction rates depicted in Fig. 2.

The enormity of the challenge for the international community to keep warming from exceeding dangerous level can be understood by the fact that the remaining carbon budget is so small, that is approximately 270 gigatons of CO_2, and current global ghg emissions are in excess of 10 gigatons per year and still rising, which means that even if the international community could stabilize global CO_2 emissions levels there would be nothing left to allocate among all nations in 23 years. This grim fact is even bleaker if the international community concludes that it should limit warming to 1.5°C, a conclusion that might become more obvious if current levels of warming start to make positive feedbacks visible in the next few years such as methane leakage from frozen tundra or more rapid loss of arctic ice.

The concept of the carbon budget explains why waiting to reduce ghg emissions levels to a certain percentage in the future is more harmful than rapid reductions earlier because the longer it takes to reduce emissions the more the remaining budget is consumed. And so the challenge for the international community to limit warming to non-catastrophic levels is not only the hard to imagine magnitude of GHG emissions reductions needed but also the speed of reductions required to live within the remaining rapidly shrinking budgets. Every day of delay in reducing GHG emissions consumes the dwindling remaining carbon budget.

4 Conclusion

The damage to the world caused by the climate change disinformation campaign is enormous. As we have seen, the campaign is responsible for an almost 50 year delay in the United States and other developed nations in taking serious steps to reduce the threat of climate change. As the world struggles to implement the 2015 Paris climate Agreement, it is already too late to prevent human induced climate change harms that are already harshly harming millions of people around the world by more intense storms, droughts which cause famine and loss of water supply needed for agriculture and drinking water, loss of glaciers which provide drinking water and water for irrigation in many parts of the world when river flows are low, the spread of tropical diseases, killer heat waves, sea level rise which is threatening cities and villages around the world. The harms already being experienced threaten hundreds of millions of the world's poorest people particularly in Africa and Southeast Asia. As devastating as these harms are to millions of climate change victims, the misery which is already visible will surely expand and intensify if the international community is not able to prevent truly catastrophic climate change which may be triggered if the international community cannot limit climate change to 1.5°C. Yet to prevent warming above 1.5°C, the international community must now reduce global GHG emissions by difficult-to-imagine rates of reduction. For these reasons, the global climate change disinformation campaign should be understood as engaging in some new kind of crime against humanity. This is so because although responsible scientific skepticism is a good thing, those responsible for the climate change disinformation campaign have been engaged in deeply morally reprehensible spread of disinformation, and in so doing are responsible for almost a 50 year delay in responding to the enormous threat of climate change.

References

Begley S (2007) Global warming deniers: a well-funded machine. Newsweek. Retrieved from http://msl1.mit.edu/furdlog/docs/2007-08-13_newsweek_global_warming_denyers.pdf

Broder J (2010) Climate change doubt is tea party article of faith. New York Times, October 21, 2009. Retrieved from http://community.nytimes.com/comments/www.nytimes.com/2010/10/21/us/politics/21climate.html?sort=newest&offset=2

Brown D (2002) American heat: ethical problems with the United States response to global warming. Rowman and Littlefield, Lantham Maryland

Brown D (2010) A new kind of crime against humanity?: The fossil fuel industry's disinformation campaign on climate change Retrieved from https://ethicsandclimate.org/2010/10/24/a_new_kind_of_vicious_crime_against_humanity_the_fossil_fuel_industrys_disinformation_campaign_on_cl/

Brown D (2011) An ethical analysis of the climate change disinformation campaign: is this a new kind of assault on humanity? Retrieved from: https://ethicsandclimate.org/2011/12/02/an_ethical_analysis_of_the_climate_change_disinformation_campaign_is_this_a_new_kind_of_assault_on_h/

Brown D (2012a) Climate change ethics: navigating the perfect moral storm. Routledge/ Earthscan, Oxon, England

Brown D (2012b) Ethical analysis of the disinformation campaign's tactics: (1) Think tanks, (2) PR campaigns, (3) Astroturf groups, and (4) Cyber-baullying attacks. Ethicsandclimate.org Retrieved from https://ethicsandclimate.org/2012/02/10/ethical_analysis_of_disinformation_campaigns_tactics_1_think_tanks_2_pr_campaigns_3_astroturf_groups/

Brulle R (2014) Institutionalizing delay in the creation of the US climate change countermovement organizations, climate change foundation funding. Clim Change 122(4)

Carbon B (2014) 2°C, the history of climate change's speed limit. Retrieved from https://www.carbonbrief.org/two-degrees-the-history-of-climate-changes-speed-limit

Dunlap RE, McCright AM (2011) Organized climate change denial. In: Dryzek JS, Norgaard RB, Schlosberg D (eds) The Oxford handbook of climate change. Oxford University Press, London, pp 144–160

Dunlap R, Brulle R (eds) (2015) Climate change and society, sociological perspectives. Oxford University Press, New York

Dunlap RE, McCright AM (2008) A widening gap: republican and democratic views on climate change. Environment magazine. Retrieved from http://www.environmentmagazine.org/archives/back%20issues/september-october2008/dunlap-full.html

Dunlap RE, McCright AM (2015) Challenging climate change, the denial countermovement. In: Dunlap R, Brulle R (eds) Climate change and society, sociological perspectives. Oxford University Press, New York

Farrell J (2015) Corporate funding and ideological polarization about climate change. Proc National Acad Sci 113(1), Retrieved from http://www.pnas.org/content/113/1/92.abstract

Hoffman A (2011) The growing climate divide. Nat Clim Change 1. Retrieved from http://webuser.bus.umich.edu/ajhoff/pub_academic/2011%20Nature%20Climate%20Change.pdf

Hoggan J (2009) Climate cover up, t*he crusade to deny global warming*.Vancouver, B.C, Greystone Books, 200

Intergovernmental Panel on Climate Change (IPCC) (2014) 5th Assessment Report, Contribution of Working Group III to the Fifth Assessment Report of the Intergovernmental Panel on Climate Change, Chapter 4. Sustainable Development and Equity. Retrieved from http://www.ipcc.ch/pdf/assessmentreport/ar5/wg3/ipcc_we3_ar5_chapter4.pdf

Jacques P, Dunlap RE, Freeman M (2008) The organization of denial: conservative think tanks and environmental scepticism. Env Polit 17:349–385

Mayer J (2016) Dark money, the hidden history of the billionaires behind the rise of the radical right. Doubleday, New York

McCright AM, Dunlap RE (2000) Challenging global warming as a social problem: an analysis of the conservative movement's counter-claims. Soc Probl 47:499–522

McCright AM, Dunlap RE (2010) Anti-reflexivity: the American conservative movement's success in undermining climate science and policy. Theory Cult Soc 26:100–133

McCright AM, Dunlap RE (2011) The politicization of climate change and polarization in the American public's views of global warming, 2001–2010. Sociol Q 52:155194

Obergassel (ne Sterk) W, Arens C, Hermwille L, Kreibich N, Mersmann F, Ott H, Wang-Helmreich H (2016). Phoenix from the ashes-An analysis of the Paris Agreement to the UNFCCC. Wuppertal Institute. Retrieved from http://wupperinst.org/uploads/tx_wupperinst/Paris_Results.pdf

Oreskes N, Conway E (2010) Merchants of doubt, how a handful of scientists obscured the truth on issues from tobacco smoke to global warming. Bloosmbury Press, New York

Pooley E (2010) Climate war, true believers, power brokers and the fight to save the earth. Hyperion, New York

Tschakert P (2015) 1.5°C or 2°C: a conduit's view from the science-policy interface at COP20 in Lima, Peru. Nature World Review. Retrieved from https://climatechangeresponses.biomedcentral.com/articles/10.1186/s40665-015-0010-z

United Nations Framework Convention on Climate Change (UNFCCC) (1992) UNFCCC/INFORMAL/84

United Nations Framework Convention on Climate Change (UNFCCC) (2015) Paris Agreement, FCCC/CP/2015/L.9/Re.1. Retrieved from https://unfccc.int/resource/docs/2015/cop21/eng/109r01.pdf

Warrick J (2015) Why are so many Americans skeptical about climate change? A study offers a surprising answer. Washington Post. Retrieved from https://www.washingtonpost.com/news/energy-environment/wp/2015/11/23/why-are-so-many-americans-skeptical-about-climate-change-a-study-offers-a-surprising-answer/?utm_term=.3c21fdd7fe9b

The Projection of Global and Regional Climate Change Models into Selected Ecosystem Functions and Services (Case Study Czech Republic)

Pavel Cudlín, Vilém Pechanec, Ondřej Cudlín, Lenka Štěrbová, and Jan Purkyt

1 Introduction

The concept of ecosystem services provides a framework for research on biodiversity and ecosystem functioning, involving the interaction of the natural environment and human activities (MEA 2005). Under changing environmental conditions the ongoing landscape fragmentation, reduction of the size of productive ecosystems and increased competition between ecosystem services are continuously increasing. Therefore, it is necessary to accumulate as much information to predict how climate and land use changes and disturbance regimes will affect the sustainability and resilience of forests and their ecosystem services. Forest-agriculture landscapes provide a broad scale of ecosystem services, e.g. production of food, wood, genetic, medicinal and decorative resources and water supply. The most important regulation services include carbon sequestration, water and soil retention, air and water purification and regulation of extreme runoff events from catchments. Aesthetic values, opportunities for recreation, and cultural and scientific inspiration are some of the most important cultural services. Although the quality of life

P. Cudlín (✉) • O. Cudlín • L. Štěrbová
Global, Change Research Institute, Academy of Sciences of the Czech Republic, České Budějovice, Czech Republic
e-mail: cudlin.p@czechglobe.cz

V. Pechanec
Department of Geoinformatics, Faculty of Science, Palacký University Olomouc, Olomouc, Czech Republic

J. Purkyt
Global, Change Research Institute, Academy of Sciences of the Czech Republic, České Budějovice, Czech Republic

Department of Plant Production and Agroecology, Faculty of Agriculture, University of South Bohemia, České Budějovice, Czech Republic

L. Westra et al. (eds.), *The Role of Integrity in the Governance of the Commons*,
DOI 10.1007/978-3-319-54392-5_9

depends on the continued provision of ecosystem services, many of them, such as climate regulation or regulation of runoff from a watershed, are non-tradeable and unpriced public goods. Quantifying the reduction in the provision of ecosystem services caused by climate change and ongoing air pollution is necessary in the near future to convince the majority of the human population that, under the precautionary principle, the best mitigation and adaptation measures should be adopted.

The aim of this chapter is to present our engagement approach as used in biodiversity valuation, which quantifies selected ecosystem functions (e.g. production) as a basis for valuation of ecosystem services (e.g. provisioning and habitat services) under conditions of climate change. We selected several climate models and created some scenarios of land cover in 2050 and 2090 using the Land Change Modeler. On the basis of these land cover scenarios, we predicted the development of carbon capture of underground and belowground parts of ecosystems. Finally, the prediction of future naturalness and biodiversity value of all habitats in the Czech Republic has been made and possible applications for mitigation and adaptation measures have been proposed.

2 Modeling Climate Change

Climate is one of the most complicated geophysical systems to simulate. Its components are as diversified as the atmosphere (air including greenhouse gases, aerosols, clouds), the hydrosphere (oceans, lakes, wetlands), the cryosphere (sea ice, land glaciers, snow), the lithosphere (land surface processes) and the biosphere (vegetation effects on the albedo, roughness and carbon cycle) (Laprise 2008). To assess the impacts of expected twenty-first century climate changes and develop corresponding mitigation and adaptation measures, regional-scale information on the nature of these changes and their uncertainties must be available. This information can be provided, for instance, by Regional Climate Models (RCMs), which are driven by Global Circulation Models (GCMs) that simulate large-scale atmospheric circulation. In 2012, a new set of GCMs was produced in the frame of the fifth phase of the Coupled Model Intercomparison Project (CMIP5), as an important background for the Fifth Assessment Report of the Intergovernmental Panel on Climate Change of the United Nations (IPPC). The CMIP5 experiment design involved different past, present and future time scales (Taylor et al. 2012) and some of the CMIP5 experiments were later downscaled to spatial scales ranging from 0.11 to 0.44 degrees over several Earth's regions by RCMs within the international CORDEX initiative (Giorgi et al. 2009). Unlike coarse resolution GCMs regional models are capable of providing more detailed information on local climate conditions and important phenomena, e.g. better representation of extreme events. The European part of CORDEX (also called as EURO-CORDEX) has recently delivered a set of high-resolution experiments including the climate scenarios till the end twenty-first century (Kotlarski et al. 2014).

Emission scenarios are an integral part of the climate models. The Intergovernmental Panel on Climate Change of the United Nations (IPCC) proposed four possible scenarios (A1, A2, B1 and B2) in 2000. In these first type of emission scenarios, different driving forces (e.g. population growth, increasing use of technology) affecting the emission load level were taken into account (Pachauri and Meyer 2014). For the current situation, a further four emission scenarios were created by the IPCC for its fifth Assessment Report (AR5) in 2014. The four scenarios, or Representative Concentration Pathways (RCPs), RCP 2.6, RCP 4.5, RCP 6.0, and RCP 8.5, are defined according to a possible range of radiative forcing values in the year 2100 relative to pre-industrial values (+2.6, +4.5, +6.0, and +8.5 W/m2, respectively) (IPCC 2014). Climate models available in many European countries have been used for regional case studies. Comparability of such information is, however, limited because different emission scenarios have been used to drive the climate models (A1B, A2), and different couplings of the Regional Climate Models and Global Circulation Models have been used to describe the future climate variability.

To examine the future climate impacts on selected ecosystem functions and services we used EURO-CORDEX simulations of three RCMs driven by different GCMs: EC-Earth_RACMO22E, MPI-ESM-LR_CLM4.8.17 and MOHC-HadGEM2-ES_RCA4 (named here GCM_RCM). All combinations GCM_RCM were combined with two standardized emission scenarios RCP 4.5 and RCP 8.5, derived by the MAGICC model (version 6). All these combinations were produced for three time horizons (i.e. 2021–2040; 2041–2060; 2081–2100); the period from 1980 to 2010 was considered as a reference period. These simulations were finally used as a source of values for successive models concerning land cover changes, biodiversity and selected ecosystem functions (Figs. 1, 2 and 3).

3 Influence of Climate Change on Land Use and Carbon Storage in a Cultural Landscape

3.1 Consequences of and Adaptations to Climate Change in Agriculture and Forestry

The combination of particularly uneven rainfall distribution, higher intensity of solar radiation and higher spring temperatures with an earlier start of the growing season is leading to the rapid exhaustion of water reserves in the soil and thus more frequent water deficits. The increased frequency of weather fluctuations (e.g. a reduced frequency of cold days and an increase in the number of warm days) is expected in the continental mid-latitudes. Extreme climatic events will significantly impact on the sectors related to water and water management, especially agriculture and forestry (Brázdil and Trnka 2015). For these reasons, it is necessary to estimate the impact of climate change on the provision of current ecosystem services.

Results from Hlavinka et al. (2015) suggest that, according to the 'dry' scenarios (such as GFCM21) crops such as barley, wheat, maize and rape (canola) will be

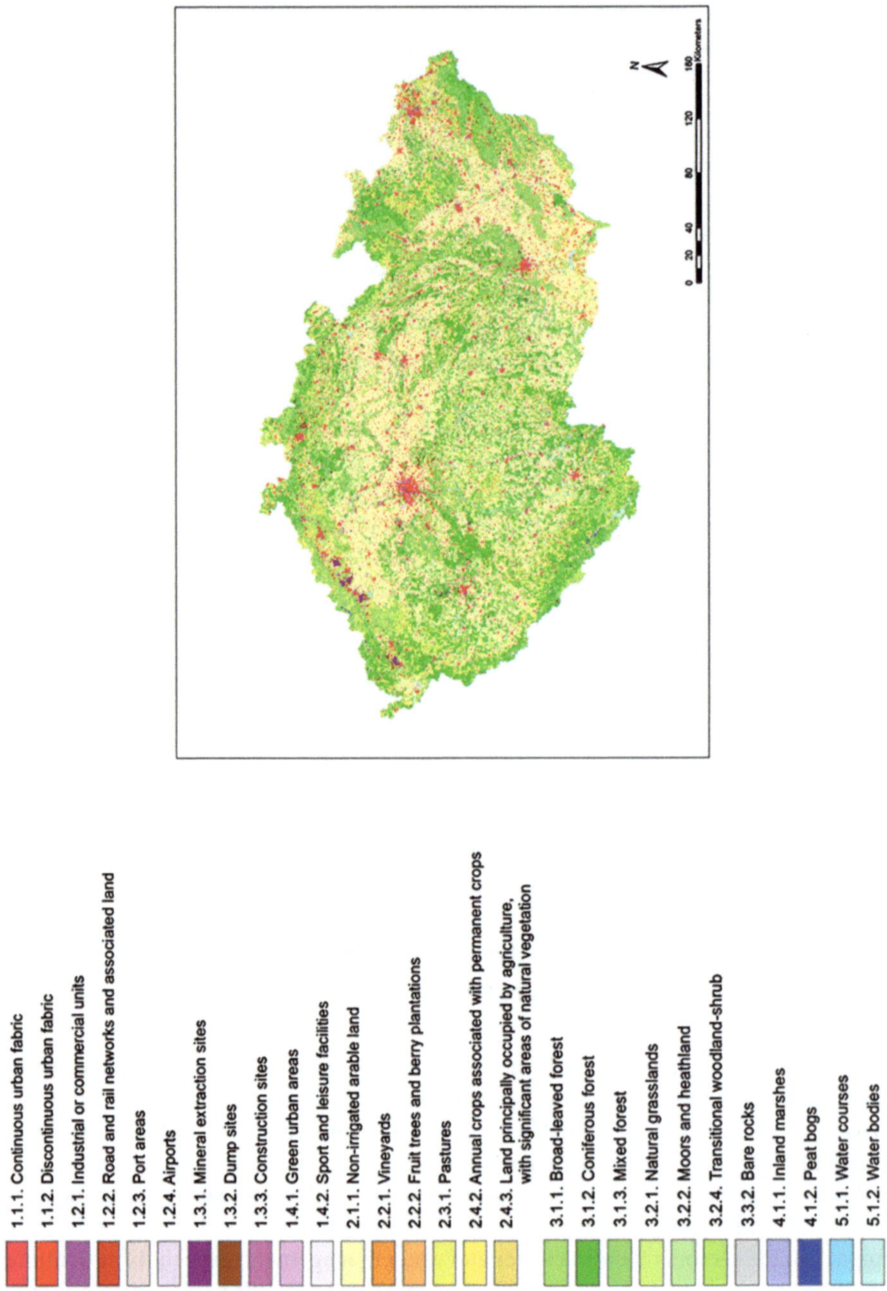

Fig. 1 Map of Land use in the year 2012

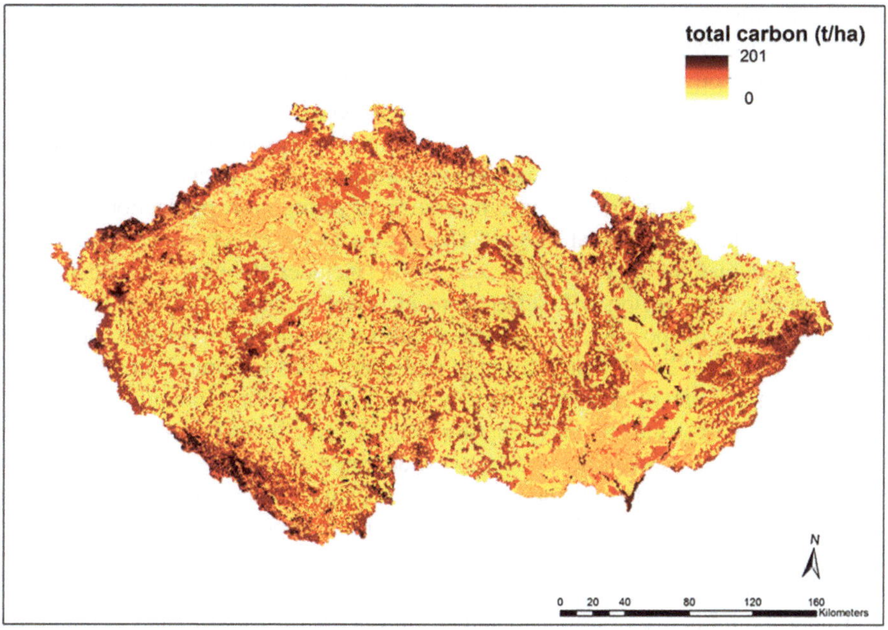

Fig. 2 The value of total carbon in the year 2012

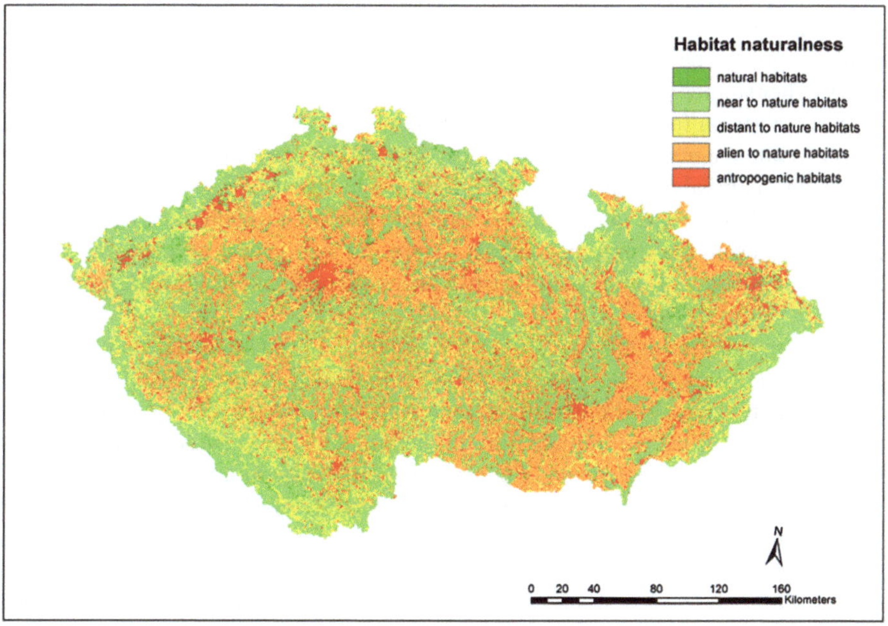

Fig. 3 Change of state of habitat naturalness by Globio3 in the year 2012

devastated for a significant number of seasons in the warmest area of the Czech Republic thus greatly reducing their future yields compared with their present yields. On the other hand, the average yield of the most important crops can be expected to increase at higher elevations (Hlavinka et al. 2015). Also, farmers will be able to take advantage of an earlier start to the growing season, at least in the lowland areas, as the proportion of days suitable for sowing increases (Trnka et al. 2011). There are also clear trends of increasing temperature affecting crop production and crop choice in southern and central Europe. The most negative effects were found for the continental climate in the Pannonian zone, which includes Hungary, Serbia, Bulgaria and Romania. This region will suffer from increased incidences of heat waves and droughts without the possibility to effectively shift crop cultivation to other parts of the year (Olesen et al. 2011). Currently proposed adaptations to climate change include changes in the timing of cultivation, variety choice, water-saving techniques, irrigation and breeding. Farmers are already adapting to a changed climate with the area of silage and grain maize expanding northwards. A wide range of adaptation options exists in most European regions to mitigate many of the negative impacts of climate change on crop production in Europe (Olesen et al. 2011). Further important adaptation measures are aimed at increasing water retention in the drought endangered agricultural landscape (Brázdil and Trnka 2015).

Under the influence of climate change, European forests will have to adapt not only to changes in mean climate variables but also to increased variability with a greater risk of extreme weather events, such as prolonged drought, storms and floods (Lindner et al. 2010). Hanewinkel et al. (2013), using the climate scenarios, estimated that between 21 and 60% of European forest lands will be suitable only for a Mediterranean oak forest type with reduced timber production and carbon sequestration properties, thus decreasing the economic returns for forest owners by 2100. Other studies predict increased global timber production due to higher growth rates and shifts in forest ranges despite likely declines in some regions (Parry et al. 2007). While the tree model of Nabuurs et al. (2002), based on ecological processes including disturbances, shows an 18% increase in stemwood growth in European forests by 2030, some other studies predict decreased production in temperate forests bordering the subtropics and an increasing significance of drought, storms, insect calamities and other major disturbances (Seppälä et al. 2009; Cudlín et al. 2013).

Modeling studies using the ALADIN-Climate/CZ climate model found that the growth conditions for Norway spruce will gradually become unsuitable in the upland and highland portions of the Czech Republic in 2040–2099. Good climatic conditions for spruce will remain only in mountainous areas, which were the areas of its original natural extension (Cudlín et al. 2012; Brázdil and Trnka 2015).

The adaptive capacity of the forest sector is relatively large in the Boreal and Temperate Oceanic regions, but is more constrained by socio-economic factors in the Temperate Continental region, and is the most limited in the Mediterranean region where large forest areas are only extensively managed or unmanaged (Lindner et al. 2010). Proposed adaptation measures include increased tree species diversity and natural regeneration, reduction of the rotation period, monitoring of diseases and pests, invasive trees and herbs, and soil climate in the forests (Brázdil and Trnka 2015).

3.2 Predictions of Changes in the Landscape

When evaluating ecosystem services, we are interested in not only the current situation, but also the development of their future provision. The prepared land use scenario is used as a basic source of information about land use/cover, which is distributed by model algorithms in the landscape; afterwards, a new rate of ecosystem service provision is computed according to this new land cover and compared in time and space. Model outputs directly display the spatial distribution and quantification of particular forms of land cover and land use. Land use/cover prediction models are based on four types of qualitative or quantitative land use scenarios: (1) projections into the future—a business as usual scenario), (2) intervention scenarios, (3) exploratory scenarios and (4) vision scenarios. Scenarios are mostly created by combining expert and participative approaches (McKenzie et al. 2012). There are many criteria that can be used to divide models, e.g. static (LUT-models) or dynamic models (e.g. Cellular automata) or to space distributed or without space distribution models. Spatially explicit approaches are often based on cellular automata, which simulate land cover changes as a function of neighbouring land cover categories and a set of drivers (Brown et al. 2013).

An appropriate set of a few models to predict land use/cover under climate change was selected. The models which include Land Change Modeler (Araya and Cabral 2010), CLUE (Verburg et al. 2010) and Urban Planner (Burian et al. 2015), have been tested in several study areas of different sizes, from small catchments to the whole Czech Republic, depending on the availability of different data sources. The Land Change Modeler (scenario type—business as usual; pixel resolution 500 m) was used to predict the development of land cover on the basis of data comparison between several historical periods (mapping Corine LC in 1990, 2000, 2006, 2012). The first step, the calculation of future land cover changes, based on the assumption of similar trend development, was assessed by a spatial-temporal analysis of changes between historical maps using Markov chains (the probability that some change will occur—where and in what direction). For modelling the allocation of the expected changes, the default methods of logistic regression and neural networks were used. The second step consisted in incorporating eight important drivers of change (altitude, slope, distance to urban area, water streams and roads, population density, predicted average daily temperature and precipitation sum and barrier of protected areas) into the land cover prediction. The output shows directly the quantity and distribution of the various forms of land use.

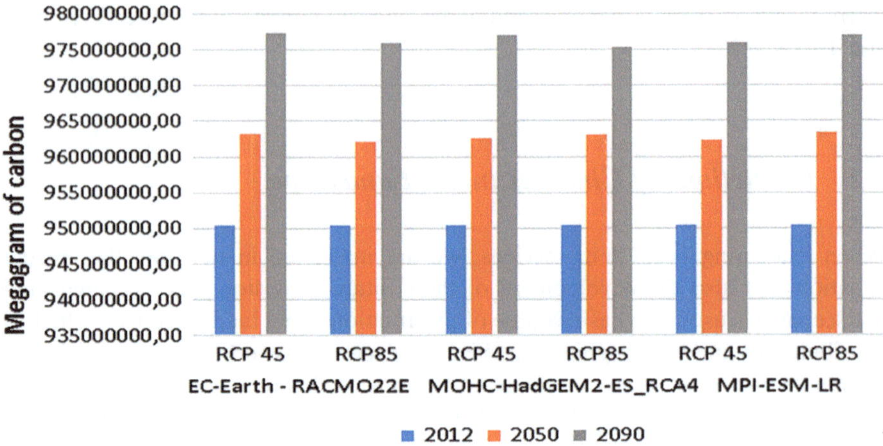

Fig. 4 Total carbon in the Czech Republic according to three climatic models and two scenarios in 2012, 2050 and 2090

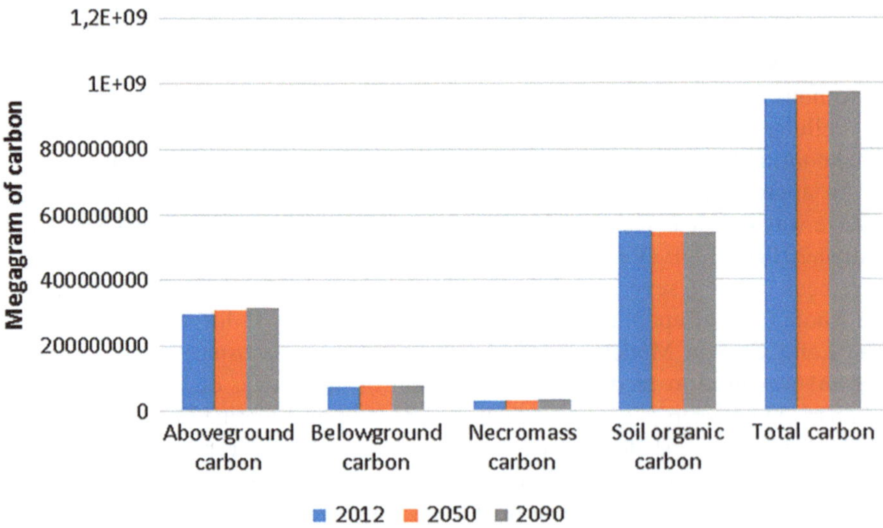

Fig. 5 Four carbon pools and total carbon in the Czech Republic in 2012, 2050 and 2090

3.3 Influences of Climate Change on Carbon Storage

Carbon storage is one of the basic parameters that informs us about the rate of provisioning ecosystem services in forestry and agriculture. Carbon storage directly reflects land use change, including land management practices and any natural or anthropogenic disturbances (Vačkář et al. 2011). Forests are the most important and up to now have increasingly been the terrestrial ecosystem sink of carbon (Erb et al. 2013; Mackey et al. 2013), due to changes in their age distribution and structure (IPCC 2000). One of the most widely used models for carbon storage in the landscape, InVEST-Carbon (Sharp et al. 2016), has been applied in many studies (e.g. Nelson et al. 2009; Nelson et al. 2010; Polasky et al. 2011; Kovacs et al. 2013; Bhagabati et al. 2014). It calculates the changes in storage for four carbon pools: above-ground biomass, below-ground biomass, necromass and soil organic carbon.

We applied the InVEST model to predict carbon storage in the Czech Republic till 2100. For each analysed landscape segment (pixel), individual carbon pools were identified based on the area of current land cover categories. In each segment, the representation of natural, near to nature and more anthropically influenced habitats was analysed. Relevant values of carbon were assigned to each pool based on the prevailing type of habitat, multiplied by the area of individual pools and finally summed for each segment. Average values of carbon for the individual pools in the land cover categories were derived from literature and our own experimental measurements (e.g. Fiala and Studený 1987; Sjörs 1991; Stará et al. 2011). A prediction of future carbon stocks was based on likely future changes in land cover, as predicted by the Land Change Modeller based on data from the mapping Corine LC. The change in carbon sequestration value was computed by subtracting the future from the current values.

According to the InVEST prediction, the amount of stored carbon will rise due to expected changes in land cover from today until 2090 (Fig. 4). Despite ever-growing discontinuous urban, fabric, industrial or commercial units, the reduction of carbon capture in these categories has not outweighed the overall increase in carbon sequestration due to a significant increase of permanent grassland and forests at the expense of arable land with a low potential to store carbon. For the individual carbon pools, the most carbon is in the soil and subsequently in the aboveground biomass (Fig. 5).

4 Influences of Climate Change on Biodiversity

Biodiversity is significantly affected by the status and changes in land use. New indicators have been developed for evaluating the current and future states of biodiversity; some of them may be included in models. Decision-makers need integrated assessment tools for natural resource planning. In regard to this need, several new concepts, tools and models were developed in the last few decades, e.g. InVEST—Integrated Valuation of Ecosystem Services and Tradeoffs (Sharp et al. 2016) and IMAGE—Integrated model to assess the global environment (MNP 2006). Model GLOBIO3 (Alkemade et al. 2009) is a good tool for assessing the current state of biodiversity while the SAR—Species Area Relationship—model (Guilhaumon et al. 2010) is useful for predicting the impact on biodiversity of climate change using climate models. Geographic Information Systems deliver large amounts of information for conservation biology by using various software packages, e.g. ArcGIS (Krigas et al. 2012) or MARXAN (Game and Grantham 2008).

Biodiversity was evaluated as Mean Species Abundance (MSA), expressed as the proportion of remaining native species from undisturbed habitats. The MSA indicator was then incorporated into the GLOBIO3 model (Alkemade et al. 2009). MSA is similar to the Living Planet Index (Loh et al. 2005) or Biodiversity Intactness Index (Scholes and Biggs 2005). MSA can be regarded as an appropriate tool for assessing the implementation of the Convention on Biological Diversity (CBD 2010) as it can be used as an indicator of the trend in the amount of selected species and for assessing the impact of human activities on biodiversity and ecosystems (UNEP 2004). The model consists of a set of regression equations describing the impact of five drivers that affect biodiversity (land use, infrastructure, landscape fragmentation, atmospheric nitrogen deposition and climate change) on biodiversity. Outputs from the model can be used as a basis for political decision making from the local to global scale, especially for governmental agencies dealing with the environment, e.g. UNEP, CBD, OECD, FAO, World Bank, EU or national governments. The GLOBIO3 model can also be used at a regional scale (e.g. assessment of the extent of biodiversity loss in Zambia (Rooij 2008)).

Another useful tool for biodiversity protection is the Marxan software, created for conservation planning. It selects areas at various scales that should be protected. These are areas with the highest biodiversity but are greatly endangered and for which protection could be obtained at the lowest total cost (Ardron 2010).

To estimate the present biodiversity status and the possible future development of that status in the Czech Republic, we divided all habitats into five groups according to their naturalness, from natural to artificial habitats, and evaluated according to biodiversity point values based on BVM—Biotope Valuation Method (Seják et al. 2003). The financial value of one point (0.58 EUR) presents the cost that Czech governments have paid in past restoration programs to increase the ecological quality of 1 m^2 of habitat by 1 point. The naturalness and values of all habitats were the basis for applying the GLOBIO3 model to estimate areas with high (hot-spots) or low biodiversity values in the future. Due the heterogeneous conditions of the Czech Republic and non-availability of a nationwide database for collecting data, instead of MSA we developed a new indicator, MHV (mean habitat value per area), which reflects the naturalness and habitat value according to BVM. Future predictions were made using the Land Change Modeler based on data from Corine LC mapping in 1990, 2000, 2006, 2012.

The most important changes in land cover and biodiversity values appears to be a greatly increasing urbanization accompanied by a considerable loss of natural and, to a lesser extent also near to nature habitats over time. Fortunately, the area of distant to nature habitats (permanent grasslands and tree monucultures) will increase at the expense of the area of alien to nature habitats (e.g. arable land). Central and Eastern Bohemia, and Central and South Moravia will be most affected by an increase in urban areas and reduction of habitat naturalness. A weak point is that the GLOBIO3 model does not consider the end of mining and reclamation of mining areas in the future and therefore it could not be applied for the mining area in Northwest Bohemia.

5 Conclusions

The outputs from the Land Change Modeler identify the quantity and distribution of the individual land cover categories and enable an estimation of derived ecosystem function performance and ecosystem service provision. Future land use development in the Czech Republic will likely show above all a decrease in arable land while urban and industrial areas increase. The outputs of the carbon pools map identify places with higher or lower carbon stocks. The change of arable land into permanent grasslands or forests is recommended to increase carbon sequestration and other water and land protection ecosystem services in the landscape.

A considerable loss of natural and near to nature habitats and an increase in anthropogenic habitats with very low biodiversity values over time were predicted by outputs of the GLOBIO3 modeling. Near to nature habitats that are not included in some category of protected areas should be at least cultivated by environmentally friendly forestry or agricultural management practices. In addition, the restoration of near to nature tree species composition, natural reservoirs of water in the form of natural wetlands in the upper catchment area and natural functions of floodplains along rivers would significantly help in adapting to climate change. This would also contribute to the planned sustainable development strategy “Strategy CR 2030”.

Acknowledgements This work is a product of the Project CzechAdapt—System for Exchange of Information on Climate Change Impacts, Vulnerability and Adaptation Measures on the Territory of the Czech Republic (EHP-CZ02-OV-1-014-2014) which was supported by grant from Iceland, Liechtenstein and Norway and by the Technological Agency of the Czech Republic No. TD03000093. This work was also supported by the Ministry of Education, Youth and Sports of CR within the National Sustainability Program I (NPU I), grant number LO1415. Habitat mapping layer provided by the Nature Conservation Agency of the Czech Republic was used.

References

Alkemade R, Oorschot MV, Miles L, Nellemann CH, Bakkenes M, Brink BT (2009) GLOBIO3: a framework to investigate options for reducing global terrestrial biodiversity loss. Ecosystems 12:374–390. doi:10.1007/s10021-009-9229-5

Araya YH, Cabral P (2010) Analysis and modeling of urban land cover change in Setúbal and Sesimbra, Portugal. Remote Sens 2:1549–1563

Ardron J (2010) Marxan good practices handbook. Available at wwwuqeduau/marxan/docs/Marxan%20Good%20Practices%20Handbook%20v2%202010pdf. Accessed 2 Nov 2016

Bhagabati NK, Ricketts T, Sulistyawan TBS, Conte M, Ennaanay D, Hadian O, McKenzie E, Olwero N, Rosenthal A, Tallis H, Wolny S (2014) Ecosystem services reinforce Sumatran tiger conservation in land use plans. Biol Conserv 169:147–156

Brázdil R, Trnka M (eds) (2015) History of weather and climate in Czech countries XI: drought in Czech countries. Past, present and future. Global Change Research Centre of the Academy of Sciences of the Czech Republic, v.v.i., Brno, p 402 (in Czech)

Brown D, Band LE, Green KO, Irwin EG, Jain A, Lambin EF, Pontius RG, Seto KC, Turner BL II, Verburg PH (2013) Advancing land change modeling: opportunities and research requirements. The National Research Council Press, Washington

Burian J, Šťastný S, Brus J, Pechanec V, Voženílek V (2015) Urban planner: model for optimal land use scenario modelling. Geografie 120:330–353

CBD (2010) Global biodiversity outlook 3. Montréal, p 94

Cudlín P, Štěpánek P, Macků J, Kohut M, Rožnovský J (2012) Influence of precipitation and temperature on the possibility to cultivate Norway Spruce in the Czech Republic in 1961–2100. In: Proceedings "Vláhové poměry krajin", Mikulov, pp 23–26 (in Czech)

Cudlín P, Seják J, Pokorný J, Albrechtová J, Bastian O, Marek M (2013) Forest ecosystem services under climate change and air pollution. In: Matyssek R, Clarke N, Cudlín P, Mikkelsen TN, Tuovinen J-P, Wiesner G, Paoletti E (eds) Climate change, air pollution and global challenges, understanding and perspectives from forest research developments in environmental science, vol 13. Elsevier, Oxford, pp 521–546

Erb KH, Kastner T, Luyssaert S, Houghton RA, Kuemmerle T, Olofsson P, Haberl H (2013) Bias in the attribution of forest carbon sinks. Nat Clim Chang 3:854–856

Fiala K, Studený V (1987) Cutting and fertilization effect on the root system in several grassland stands. Ekológia 6:389–402

Game E, Grantham H (2008) Marxan user manual. University of Queensland. Available at http://marxannet/userguides. Accessed 2 Nov 2016

Giorgi F, Jones C, Asrar GR (2009) Addressing climate information needs at the regional level: the CORDEX framework. WMO Bull 58:175–183

Guilhaumon F, Mouillot D, Gimenez O (2010) mmSAR: an R-package for multimodel species-area relationship inference. Ecography 33:420–424

Hanewinkel M, Cullmann DA, Schelhaas MJ, Nabuurs GJ, Zimmermann NE (2013) Climate change may cause severe loss in the economic value of European forest land. Nat Clim Chang:203–207. doi:10.1038/NCLIMATE1687

Hlavinka P, Kersebaum KC, Dubrovský M, Fischer M, Pohanková E, Balek J, Žalud Z, Trnka M (2015) Water balance, drought stress and yields for field crop rotations under present and future conditions in the Czech Republic. Clim Res 65:175–192. doi:10.3354/cr01339

IPCC (2000) Land Use, land-use change, and forestry. Cambridge University Press, Cambridge, p 375

IPCC (2014) Climate change 2014: impacts, adaptation, and vulnerability. Cambridge University Press, Cambridge/New York

Kotlarski S, Keuler K, Christensen OB, Colette A, Déqué M, Gobiet A, Goergen K, Jacob D, Lüthi D, van Meijgaard E, Nikulin G, Schär C, Teichmann C, Vataurd R, Warrach-Sagi K, Vulfmeyer V (2014) Regional climate modeling on European scales: a joint standard evaluation of the EURO-CORDEX RCM ensemble. Geosci Mod Dev 7:1297–1333

Kovacs K, Polasky S, Nelson E, Keeler BL, Pennington D, Plantinga AJ, Taff SJ (2013) Evaluating the return in ecosystem services from investment in public land acquisitions. PLoS One 8(6):e62202

Krigas N, Papadimitriou K, Mazaris, AD (2012) GIS and ex situ plant conservation. In: Alam BM (ed) Application of geographic information systems. INTECH Open Access Publisher

Laprise R (2008) Regional climate modelling. J Comput Phys 227:3641–3666

Lindner M, Maroschek M, Netherer S, Kremer A, Barbati A, Garcia-Gonzalo J, Seidl R, Delzon S, Dodona P, Kostro M, Lexer MJ, Marchetti M (2010) Climate change impacts, adaptive capacity, and vulnerability of European forest ecosystems. For Ecol Manag 259:698–709

Loh J, Green RE, Ricketts T, Lamoreux J, Jenkins M, Kapos V, Randers J (2005) The living planet index: using species population time series to track trends in biodiversity philosophical transactions of the royal society. Biol Sci 360(1454):289–295

Mackey B, Prentice IC, Steffen W, House JI, Lindenmayer D, Keith H, Berry S (2013) Untangling the confusion around land carbon science and climate change mitigation policy. Nat Clim Chang 3(6):552–557

McKenzie E, Rosenthal A, Bernhardt J, Girvetz E, Kovacs K, Olwero N, Toft J (2012) Developing scenarios to assess ecosystem service tradeoffs: guidance and case studies for InVEST users. Natural capital project. Available at http://wwwnaturalcapitalprojectorg/pubs/ScenariosGuidepdf. Accessed 2 Nov 2016

MEA (2005) Ecosystems and human well-being: wetlands and water synthesis. Millennium ecosystem assessment. World Resources Institute, Washington, DC

MNP (2006) Integrated modelling of global environmental change: an overview of IMAGE 24. Netherlands Environmental Assessment Agency Bilthoven, The Netherlands

Nabuurs GJ, Pussinen A, Karjalainen T, Erhard M, Kramer K (2002) Stemwood volume increment changes in European forests due to climate change – a simulation study with the EFISCEN model. Glob Chang Biol 8:304–316

Nelson E, Mendoza G, Regetz J, Polasky S, Tallis H, Cameron D, Chan KM, Daily GC, Goldstein J, Kareiva PM, Lonsdorf E, Naidoo R, Ricketts TH, Shaw M (2009) Modeling multiple ecosystem services, biodiversity conservation, commodity production, and tradeoffs at landscape scales. Front Ecol Environ 7:4–11

Nelson E, Sander H, Hawthorne P, Conte M, Ennaanay D, Wolny S, Manson S, Polasky S (2010) Projecting global land-use change and its effect on ecosystem service provision and biodiversity with simple models. PLoS One 5(12):e14327

Olesen JE, Trnka M, Kersebaum KC, Skjelvåg AO, Seguin B, Peltonen-Sainio P, Rossi F, Kozyra J, Micale F (2011) Impacts and adaptation of European crop production systems to climate change. Eur J Agron 34:96–112

Pachauri RK, Meyer L (eds) (2014) Climate change, synthesis report. A report of the Intergovernmental Panel in Climate Change, p 169

Parry ML, Canziani OF, Palutikof JP, van der Linden PJ, Hansosn CE (eds) (2007) Climate change *2007:* impacts, adaptation and vulnerability. Contribution of Working Group II to the fourth assessment report of the intergovernmental panel on climate change, Cambridge University Press, Cambridge, p 1000

Polasky S, Nelson E, Pennington D, Johnson KA (2011) The impact of land-use change on ecosystem services, biodiversity and returns to landowners: a case study in the state of Minnesota. Environ Resour Econ 48(2):219–242

Rooij W (2008) Manual for biodiversity modelling on a national scale using GLOBIO3 and CLUE methodology to calculate current and future status of biodiversity. Case study area: Zambia. MNP Bilthoven, Netherlands, p 25

Scholes RJ, Biggs R (2005) A biodiversity intactness index. Nature 434(7029):45–49

Seják J, Dejmal I, Petříček V, Cudlín P, Míchal I, Černý K, Kučera T, Vyskot I, Strejček J, Cudlínová E, Cabrnoch J, Šindler M, Prokopová M, Kovář J, Kupka M, Ščasný M, Šafařík M, Roušalová S, Stejskal V, Zapletal J (2003) Valuation and prizing of biotopes of the Czech Republic. Czech Ecological Institute, p 422 (In Czech)

Seppälä R, Buck A, Katila P (eds) (2009) Adaptation of forests and people to climate change. A global assessment report IUFRO World Series Volume 22, Helsinki, p 224

Sharp R, Tallis, HT, Ricketts T, Guerry AD, Wood SA, Chaplin-Kramer R, Nelson E, Ennaanay D, Wolny S, Olwero N, Vigerstol K, Pennington D, Mendoza G, Aukema J, Foster J, Forrest J, Cameron D, Arkema K, Lonsdorf E, Kennedy C, Verutes G, Kim CK, Guannel G, Papenfus M, Toft J, Marsik M, Bernhardt J, Griffin R, Glowinski K, Chaumont N, Perelman A, Lacayo M, Mandle L, Hamel P, Vogl AL, Rogers L, Bierbower W (2016) InVEST +VERSION+ user's guide. The Natural Capital Project, Stanford University, University of Minnesota, The Nature Conservancy, and World Wildlife Fund

Sjörs H (1991) Phyto- and Necromass above and below Ground in a Fen. Holarct Ecol 14:208–218

Stará L, Matějka K, Cudlín P, Bodlák L, Pokorný J, Středa T, Čížková H, Pechar L, Burešová R, Zemek F, Marek MV (2011) Carbon storage in the vegetation of the Czech Republic and model

landscape carbon budget. In: Marek M (ed) Carbon in ecosystems of the Czech Republic under climate change. Academia, Praha, pp 189–210 (In Czech)
Taylor KE, Stouffer RJ, Meehl GA (2012) An overview of CMIP5 and the experiment design. Am Meteorol Soc 93:485–498
Trnka M, Eitzinger J, Semerádová D, Hlavinka P, Balek J, Dubrovský M, Kubu G, Štěpánek P, Thaler S, Možný M, Žalud Z (2011) Expected changes in agroclimatic conditions in Central Europe. Clim Chang 108:261–289
UNEP (2004) Decision VII/30 strategic plan: future evaluation of progress. In: Seventh conference of the parties to the convention on biological diversity, Kuala Lumpur
Vačkář D, Melichar J, Lorencová E (2011) Ecosystem services of grasslands in the Czech Republic. Report to the agency for nature conservation and landscape protection of the Czech Republic and European Topic Centre on Biological Diversity, p 75
Verburg P, Veldkamp T, Lesschen JP (2010) Exercises for the CLUE-S model. University Amsterdam. http://citeseerx.ist.psu.edu/viewdoc/download?doi=10.1.1.521.1688&rep=rep1&type=pdf. Accessed 2 Nov 2016

Bio-economy as a New Perspective for Solving Climate Change?

Eva Cudlínová, Miloslav Lapka, and Jan Vávra

The twenty-first century has been heralded as the 'century of biology' and the bio-sciences have been compared with the physical sciences in terms of their potential to 'transform society' (National Research Council 2009; Petersen and Krisjansen 2015).

1 Introduction

According to the latest Intergovernmental Panel on Climate Change (IPCC) report, 65% of overall emissions of greenhouse gases (GHG) released in 2010 were emissions of CO_2 from fossil fuels and industrial processes (IPCC 2014). As Scarlat et al. (2015) summarized, the EU set many targets and introduced many policies to lower the emissions of greenhouse gases while being a competitive and innovative economy. In 2007, the European Council released the 2020 Climate and Energy Package which aimed at a 20% reduction in GHG emissions (compared to 1990) plus that 20% of energy would come from renewable sources and a 20% improvement in energy efficiency (compared to a business as usual scenario). All of these had to be fulfilled by 2020. These goals were updated in 2014 and new goals were set for 2030: a 40% cut of emissions, 27% of energy from renewables and 27% of energy efficiency improvement. The EU roadmap for a low carbon economy in 2050 suggests scenarios for massive decarbonization by 2050, aiming to achieve a 80% decrease of overall GHG emissions (compared to 1990) and almost the total decarbonization of the energy sector (European Union 2011). Such ambitious goals were supported by the COP 21 meeting in Paris 2015 and the subsequent Paris Agreement. Countries there agreed to keep the increase of average global temperatures under 2 °C above the pre-industrial level (European Commission 2016a).

E. Cudlínová (✉) • M. Lapka • J. Vávra
Department of Regional Management, Faculty of Economics, University of South Bohemia in České Budějovice, České Budějovice, Czech Republic
e-mail: evacu@ef.jcu.cz; mlapka@ef.jcu.cz; jvavra@ef.jcu.cz

L. Westra et al. (eds.), *The Role of Integrity in the Governance of the Commons*,
DOI 10.1007/978-3-319-54392-5_10

The bio-economy, which is based on bio-energy produced from biomass and waste opens up the possibility to escape from coal and oil resource dependence. Bio-energy is expected to have a crucial role among renewables by 2020. According to an European Environment Agency study, biomass could account for up to 236 million tons of equivalent oil out of a total of 1.8 billion tons of primary energy requirement for the whole EU in 2020 (European Environment Agency 2006). The more recent report by European Commission concludes that various scenarios would range between 190 and 230 million tons (European Commission 2016b). These numbers make up about 10–13% of primary energy demand.

Replacing fossil fuels with biomass could lead to a decrease of CO_2 emissions and help solve the problem of global climate change. In other words, the bio-economy may be a way of decoupling economic growth from environmental degradation. This chapter attempts to explore the promises and real possibilities of the bio-economy in fulfilling this expectation focusing on some serious risks in the case of its sustainable global application.

2 Bio-economy in Political Documents and Scientific Literature

2.1 *Bio-economy in Political Documents*

Bio-economy is a relatively new phenomenon which appears in many political documents and strategies. Recently published policy agendas for the bio-economy by the Organization for Economic Cooperation and Development (OECD), the European Union (EU) and the United States of America (US), suggest that significant advances and support can be expected in the field over the coming decades at regional, national and international levels (The White House 2012; McCormick and Kautto 2013).

The term bio-economy has its origin in life sciences and biotechnology spheres. It was first defined by Juan Enriquez, Andrew Cabot and Rodrigo Martinez in 1997 (Enriquez et al. 1998). These two life science business innovators position themselves as pioneer entrepreneurs who offer prophecies about bio-enabled futures. Their original concept has been extended to incorporate other ideas such as the bio-refinery concept. At present, the bio-economy penetrates into whole branches and sectors of the economy and may be compared with information and technological revolution (ITC) or revolution 4.0 (von Braun 2013). Bio-economy may also be seen as an old practice that has existed in societies for a long time as practices of food production such as a bakery or brewery, etc.

Apart from the political support of bio-economy introduction into the present economic systems, there is not any unified policy definition of what bio-economy really is. The OECD states that "the bio-economy involves three elements: the use of advanced knowledge of genes and cell processes to design and develop new processes and products; the use of renewable biomass and efficient bio-processes to stimulate sustainable production; and integration of biotechnology knowledge and applications across a range of sectors" (OECD 2009). In its policy packages, the European Commission (EC) defines bio-economy as "the production of renewable biological resources and the conversion of these resources and waste streams into value added products, such as food, feed, bio-based products and bio-energy" (European Union 2012). In addition, the EC background paper on the public consultation determines that "a bio-based economy integrates the full range of natural and renewable biological resources–land and sea resources, biodiversity and biological materials (plant, animal and microbial), through to the processing and the consumption of these bio-resources. The bio-economy encompasses the agriculture, forestry, fisheries, food and biotechnology sectors, as well as a wide range of industrial sectors, ranging from the production of energy and chemicals to building and transport" (European Commission 2011).

In turn, the USA bio-economy blueprint, which was published soon after the EC policy package in 2012, determines that "—the bio-economy is one based on the use of research and innovation in the biological sciences to create economic activity and public benefit" (The White House 2012). All these political definitions give us a broad space for interpretation in the case of the application for economic support.

2.2 *Bio-economy in Scientific Literature*

The broad range of political definitions reflects the situation in the research field where the precise scientific definition of bio-economy is also missing. It is not clear whether it is a new branch of economics, a new economic paradigm or just a specific part of the green economy based on biological resources and the application of biotechnology (Maciejczak 2015).

In both political documents and scientific literature we can find at least three different names: bio-economy, bio-based economy or knowledge based bio-economy, which are used as synonyms even if they differ in their real meaning in national context. For example, the 'bio-based economy' tends to focus on 'raw materials', namely, natural and renewable biological resources, while 'bio-economy', on the other hand, tends to be used to designate biotechnology, life science and other related technologies (Staffas et al. 2013).

The main advantage of the bio-economy comes not only from the use of renewable resources such as the green economy does, but also from the specific use of biological resources, including the biomass that may be produced in fields, forests and oceans. Therefore, human activities such as agriculture, forestry or fishery, can help in producing biomass. The basic building blocks for materials, chemicals and energy are derived from renewable biological resources such as plant and animal sources. An economy founded on biomass instead of fossil fuels represents a significant shift in socio-economic, agricultural, energy and technological systems. We can talk about the biologization of economy.

Furthermore, an additional value of the bio-economy is inherent in the reusing of resources and the enormous capacity for waste recycling in so the called cascading use of resources. There are many researchers who support this new economic way or perspective (Asveld et al. 2011; Kornerup Bang et al. 2009; Paula and Birrer 2006; Soetaert and Vandamme 2006).

It seems that there is no place for doubts or critique but, in fact, the principle of bio-economy is a controversial phenomenon. From the sustainability point of view, issues of scale, time and human consumption patterns must be take into account. Special attention should be paid to the scale of application, e.g., if the bio-economy is presented as a global solution or regional strategy.

3 Main Risks of Bio-economy Implementation on a Global Scale

3.1 Bio-economy and Carbon Neutrality

The main promise of the bio-economy is based on the reduction of the negative effects of climate change. This idea comes from the predisposition that replacing fossil fuels with biological resources will result in decreasing CO_2 emission. Given the fact that fossil fuel use is the primary source of CO_2, this sounds reasonable. Unfortunately, the way in which people use land is not CO_2 neutral as well.

The idea of biomass production as a main way of CO_2 mitigation is not unanimously supported by the research community. In the scientific literature we can find a group of pessimists regarding the potential of agriculture to solve the problems with CO_2 emissions (IPCC 2001; Paustian et al. 2004). According to the latest research reports, agriculture as well as other industrial branches emit significant amounts of CO_2, CH_4, and N_2O into the atmosphere (IPCC 2001). CO_2 is released largely through microbial decay or the burning of plant litter and organic soil matter (Smith et al. 2007).

The Intergovernmental Panel on Climate Change (IPCC) reported that current agricultural land use and land conversion contributions to GHG emissions (CO_2, CH_4, and N_2O) are globally estimated to be at least 2.5 times greater than the total emissions from global transport. Agriculture CO_2 emissions is nearly at the same level as those caused by electricity and heat production. The situation is more complicated with the so called indirect land use changes (ILUC). This refers to the clearing of forests and of putting pastures into agricultural use and as a result of the conversion of crop land into the production of biomass. Quantifying these emissions remains a great challenge. Greenhouse gas emissions from this sector come mostly from the cultivation of crops and livestock, and deforestation.

This relatively pessimistic accounting as regards agriculture and its emissions suffers from some difficulties. The main weakness of this CO_2 emission calculation comes from the fact that this estimate does not include the phenomenon of CO_2

sequestration. In other words, the amount of CO_2 that ecosystems remove from the atmosphere during their lifecycle and which represents an offset of approximately 20% of overall emissions. After including this offset, the real agricultural emissions fall down from 24 to 4%, which is much a better balance in favor of agriculture (EPA 2016).

Even if the sequestration effect of soil, plants and forests has an enormous importance, it does not mean that replacing fossil fuels with bio-energy will reduce carbon emissions by itself. The CO_2 released by tail pipes and smokestacks is roughly the same per unit of energy regardless of the source. "Bio-energy reduces greenhouse emissions only if the growth and harvesting of the biomass for energy captures carbon *above and beyond* what would be sequestered anyway, and offsets the emissions from energy use. For example, if unproductive land supports fast-growing grasses for bio-energy, or if forestry improvements increase tree growth rates, the additional carbon absorbed offsets emissions when burned for energy. Energy use of manure or crop and timber residues may also capture additional carbon" (ISIS 2010).

The danger of false carbon accounting methods for bio-fuels that ignores emissions in land-use changes is the distortion of facts enabling natural habitat destruction. As a most serious example could serve the process of deforestation caused by the replacement of natural forest for bio-fuels production and industrial tree plantations. The direct link between deforestation and bio-fuels occurs when forests are cleared to establish bio-fuel crops. The indirect link is when bio-fuels production moves on to croplands or pastures, and leads to new forest clearings to relocate agriculture (Searchinger et al. 2008). The lack of distinction between natural forests and tree plantations (for example, in the FAO formal definition of forest), results in additional incentives to replace natural forests with fast growing tree plantations, including exotic species, for pulp and biomass production (Melillo et al. 2009).

If bio-energy crops displace forests or grasslands, the carbon released from soils and vegetation, plus lost future sequestration, generates a carbon debt which counts against the carbon the crops absorb. In this case it is not just increasing CO_2 but also raising the pressure on limited land resources and increases competition between bio-fuels and food production and biodiversity.

3.2 *Availability of Land for the Bio-economy*

Biomass is the fourth largest energy source after coal, oil and natural gas. However, a crucial question remains to be answered: "how much of it could be produced sustainably"? When thinking about sustainability of the bio-economy, it is necessary to be aware of the amount of land which is available for such activity. In the case of agricultural land, the absolute area matters as well as its geographical placement. Several studies have tried to estimate the available amount of biomass in Europe and worldwide, now and into the future. All of these studies show large uncertainties. Europe is an interesting case because it has limited land available for growing biomass in comparison with its ambitious bio-economy plans. The total supply of sustainable biomass in 2030 may not be enough to fulfill the demand of a 10% bio-based economy (Ros et al. 2012).

The conditions of bio-based production in the EU are even more serious from the perspective of a footprint calculation. According to the land footprint indicator, the EU is a "net importer" of cropland (Bringezu et al. 2012). In comparison to the amount of global cropland available on a per capita basis, the EU used around one-third more than the global average in 2007. This raises questions about the sustainability of pursuing a bio-economy strategy and whether it would further raise the EU's global land requirements (O'Brien et al. 2015).

The US position is more favorable if compared to Europe. A report by the US Department of Energy assumes that the availability of biomass resources are capable of substituting 30% of the nation's petroleum consumption (OECD 2014).

Most assessments reveal major potential about 80% of the world's fuel wood supply are situated in Africa, Asia and South America, and in regions that are distant from biomass markets and demand (Lewandowski 2015).

Under such conditions the biomass could be traded internationally and the EU would be one region that would probably depend on the world market to supply its bio-economy with biomass in the future. The main source of biomass would most likely come from the regions mentioned above which have low food security and where agricultural production is often not considered "sustainable" (e.g., due to deforestation to expand agricultural production). Such a scenario is highly controversial (Pavanan et al. 2013).

Apart from this imbalanced placement of biomass supply presented, political estimations of total biomass availability are grossly overestimated. Analyses of potential revealed that about 250–1580 Mha of marginal, degraded or abandoned agricultural land could be available for biomass production (Dauber et al. 2012). The figures depend on the definition of "marginal land"—which categories (waste land, degraded land, abandoned land, etc.) it encompasses (Lewis and Kelly 2014). References to large areas of available "marginal lands" are often fictional because they are based on a devaluation of the many uses of land by indigenous peoples, peasant farmers, pastoralists, as well as and for biodiversity, water and soil protection. "Abandoned cropland" also includes large areas of land where tropical forests were destroyed for plantations and cattle ranching and where soil degradation and water depletion now make agriculture difficult.

We could discuss how precise are the estimates of available land, but the phenomenon of "land grabs" market speculation and investment in land around the world to gain access to lands, gives evidence that there exists strong competition for land-use. Recent research from the International Land Coalition indicates about 44% of land grabs have been for the purpose of growing bio-energy crops (Biofuelwatch et al. 2011). According to De Schutter (2011), land grabbing (often happening in Africa and Asia) may reduce the access of local farming communities to land and water resources, thus raising land prices with negative effects on local development options and food security.

4 Sustainable Way for the Bio-economy

4.1 Sustainable Bio-economy and the Circular Economy

The wide diversity of positions on the sustainability of biomass supply is grounded in the different and often contradictory interests and perceptions of the stakeholders involved. In the sustainable or "ideal" bio-economy, biomass production will take ecological, social and health aspects into consideration and be internationally competitive (Lewandowski 2015).

To be in accordance with sustainability, a broad range of criteria need to be addressed—soil conservation issues and food security as well as biodiversity preservation. "Whereas certification schemes offer concrete rules, guidelines and implementation strategies for most criteria for sustainable biomass production at local and project level, they cannot provide effective strategies for dealing with the competing claims on biomass or on land used for its production. These include the "food-versus-fuel" as well as the "food, energy, and environment trilemma" (Pfau et al. 2014).

The principle of circular economy based on the near absolute recycling and cascading use of resources seems to be very close to the sustainable principle. Indeed, it could serve as a synonym to sustainable bio-economy. The circular economy goes beyond recycling as it is based on a restorative industrial system geared towards the designing out of waste, not just designing for a better end of-life recovery but to minimize energy use as well.

The notion of a circular economy was first touted in the 1970s by environmental academics John T. Lyle and Walter Stahel, but only really caught on when former sailor Dame Ellen MacArthur set up the Ellen MacArthur Foundation in 2010 to champion the concept (Perella 2016).

Contrary to the Environmental Kuznets Curve (EKC), environmental quality cannot be maintained or improved via economic growth. Instead, the improvement in environmental quality, as measured by a reduction in pollution, can only be achieved by an increase in the environmental self-renewal rate or the recycling ratio (George et al. 2015).

The most sustainable way of circular economy (see Fig. 1) as a form of bio-economy relies on the use of waste as main source materials. In this case the circular and cascading form of bio-economy is fully realized.

One of the hurdles that the circular economy faces is that it is usually more expensive to manufacture a durable long lasting good than an equivalent quick and disposable version (Ellen MacArthur Foundation 2016). That is the main reason why the scaling up of a circular economy at an international level will likely require government support. Some countries are already starting to act—for example China has set up CACE, a government-backed association to encourage circular growth and the EU has also formally supported the circular economy.

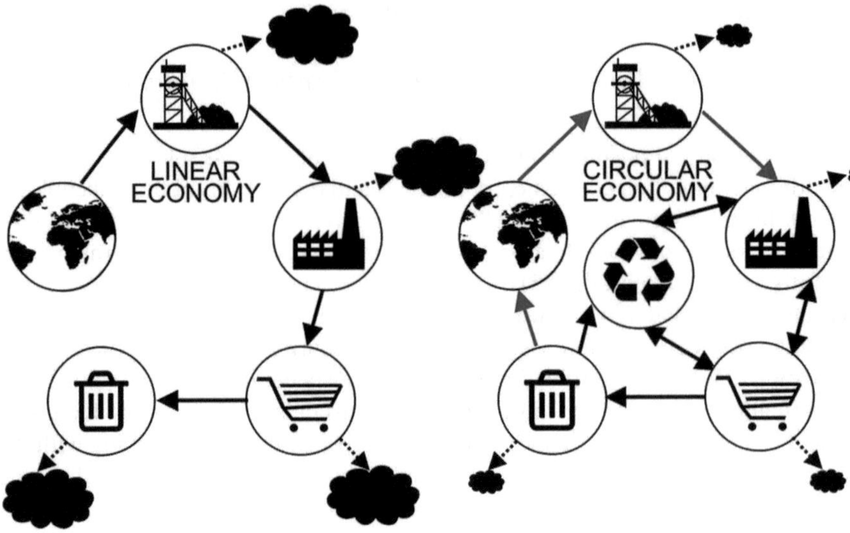

Fig. 1 Scheme of circular economy. Source: Sauvé et al. (2016)

4.2 Regional Dimension of Bio-economy: Responsible Bio-economy, Public Goods

The question of scale is a crucial predisposition of sustainability—the regional dimension represents a more sustainable way for bio-economy application. The use of local resources with a combination of local knowledge enables feedbacks between the economy, the environment and the local community. Farmers play an important role as representatives of local knowledge (Schmid et al. 2012). Von Braun (2013) identified the lack of effective land-use planning for biomass production without the involvement of small holder farmers. Rural regions represent a basic background for the bio-economy as natural producers of biomass resources.

On the other hand, the bio-economy as a new way of production could open a new perspective, especially for marginal, less developed rural regions with dispersed populations (Lehtonen and Okkonen 2016). A bio-economy is well suited to peripheral areas because the industries entering new regions tend to relate to pre-existing industries. However, new production activities such as bio-economy could attract new investors without any relationship to former businesses.

In the petro-economy, most rural areas (especially agricultural areas) use more energy than they produce and rising energy costs hurt them. Bio-economy practice will enable these areas to produce more of their own energy sources than they can consume. From this perspective, rural regions become the beneficiaries (Brown 2003). The great advantage could be the circular character of local economy in which most of the produced energy will not need to be transported out and in. Rural regions will save both in terms of energy and transport costs (Johnson and Altman 2014).

In the bio-economy we will replace oil (as well as coal and natural gas) with biomass-based material. Therefore, in the bio-economy, the agricultural economic issues will be much the same as in the petro-economy.

The way for sustainable living cannot be found without changing our consumption patterns as was mentioned in Agenda 21, chapter 4.1. (United Nations 1992) Agenda 21 "contains the following program areas: (a) Focusing on unsustainable patterns of production and consumption; (b) Developing national policies and strategies to encourage changes in unsustainable consumption patterns." Of the same importance as the changing the consumption patterns is the problem of solving the current imbalances in the global patterns of consumption and production.

5 Conclusion

Our chapter highlighted some of the main risks and weak points connected with the global application of bio-economy as a general solution to our present socio-economic and environmental challenges. We do agree with the opinion that there is not sustainable or unsustainable biomass, but more precisely sustainable or unsustainable use of it. A similar situation is in the case of the sustainability of the bio-economy. Whatever definition of bio-economy we decide to use (EU, US or OECD), we have free space for our own interpretation. The most important predisposition for the sustainable use of bio-economy is the scale of bio-economy application. This focuses on patterns of consumption and preference for using all types of waste as a main source of bio-economy production instead of only the biomass competing with food production.

References

Asveld L, Van Est R, Stemerding D (2011) The bio-economy: fertile soil for policy targets. In: Asveld L, Van Est R, Stemerding D (eds) Getting to the core of the bio-economy: a perspective on the sustainable promise of biomass. Rathenau Instituut, The Hague

Biofuelwatch, Global Forest Coalition, Global Justice Ecology Project, EcoNexus, Biomass Accountability Project, Partnership for Policy Integrity, PT AirWatchers (Port Townsend, Washington) and Ozark Riverkeepers Network (2011) Submission to Rio Plus 20 zero draft: a green economy" cannot run on biofuels or bioenergy. www.biofuelwatch.org.uk/docs/Bioenergy-for-Rio-zero-draft-1.pdf

Bringezu S, O'Brien M, Schütz H (2012) Beyond biofuels: assessing global land use for domestic consumption of biomass. A conceptual and empirical contribution to sustainable management of global resources. Land Use Policy 29(1):224–232

Brown RC (2003) Biorenewable resources: engineering new products from agriculture. Iowa State Press, Ames

Dauber J, Brown C, Fernando AL, Finnan J, Krasuska E, Ponitka J, Styles D, Thraen D, Van Groenigen KJ, Weih M, Zah R (2012) Bioenergy from "surplus" land: environmental and socio-economic implications. BioRisk 7:5–50

De Schutter O (2011) How not to think of land grabbing: three critiques of largescale investment in farmland. J Peasant Stud 38(2):249–279

Ellen MacArthur Foundation (2016) Circular economy. www.ellenmacarthurfoundation.org/circular-economy. Accessed 10 Oct 2016

Enriquez J, Martinez R, Cabot A (1998) Biotechnology in Europe. Ingenioren, Denmark, p 14, August 5 1998

EPA (2016) Global greenhouse gas emissions data. US Environmental Protection Agency. www.epa.gov/ghgemissions/global-greenhouse-gas-emissions-data. Accessed 8 Oct 2016

European Commission (2011) Bio-based economy in Europe: state of play and future potential - Part 2 Summary of the position papers received in response of the European Commission's Public on-line consultation. ec.europa.eu/research/consultations/bioeconomy/bio-based-economy-for-europe-part2.pdf. Accessed 8 Oct 2016

European Commission (2016a) Climate action: Paris Agreement. ec.europa.eu/clima/policies/international/negotiations/paris/index_en.htm. Accessed 15 Oct 2016

European Commission (2016b) Agriculture and rural development: biomass potential. ec.europa.eu/agriculture/bioenergy/potential/index_en.htm. Accessed 15 Oct 2016

European Environment Agency (2006) How much bioenergy can Europe produce without harming the environment? EEA Report No 7/2006. www.eea.europa.eu/publications/eea_report_2006_7/at_download/file. Accessed 8 Oct 2016

European Union (2011) A roadmap for moving to a competitive low carbon economy in 2050, COM/2011/112 final. eur-lex.europa.eu/LexUriServ/LexUriServ.do?uri=COM:2011:0112:FIN:en:PDF. Accessed 8 Oct 2016

European Union (2012) Innovating for sustainable growth: a bioeconomy for Europe, COM/2012/060 final. eur-lex.europa.eu/legal-content/EN/TXT/?uri=CELEX%3A52012DC0060. Accessed 8 Oct 2016

George DAR, Chi-ang Lin B, Chen Y (2015) A circular economy model of economic growth. Environ Model Softw 73:60–63

IPCC (2001) In: Houghton JT, Ding Y, Griggs DJ, Noguer M, van der Linden PJ, Dai X, Maskell K, Johnson CA (eds) Climate change 2001: the scientific basis. Contribution of Working Group I to the third assessment report of the intergovernmental panel on climate change, Cambridge University Press, Cambridge

IPCC (2014) Climate change 2014 synthesis report summary for policymaker. www.ipcc.ch/pdf/assessment-report/ar5/syr/AR5_SYR_FINAL_SPM.pdf

ISIS (2010) Scientists expose devastating false carbon accounting for biofuels. Institute of Science in Society ISIS, Report 29/11/10. www.i-sis.org.uk/falseAccountingForBiofuels.php. Accessed 8 Oct 2016

Johnson TG, Altman I (2014) Rural development opportunities in the bioeconomy. Biomass Bioenergy 63:341–344

Kornerup Bang J, Follér A, Buttazzoni M (2009) Industrial biotechnology. More than green fuel in a dirty economy? Exploring the transformative potential of industrial biotechnology on the way to a green economy. WWF Denmark, Copenhagen

Lehtonen O, Okkonen L (2016) Socio-economic impacts of a local bioenergy-based development strategy – the case of Pielinen Karelia. Renew Energy 85:610–619

Lewandowski I (2015) Securing a sustainable biomass supply in a growing bioeconomy. Glob Food Sec 6:34–42

Lewis SM, Kelly M (2014) Mapping the potential for biofuel production on marginal lands: differences in definitions, data and models across scales. Int J Geo-Inf 3(2):430–459

Maciejczak M (2015) How to analyze bioeconomy? Roczniki (Annals), issue 6

McCormick K, Kautto N (2013) The bioeconomy in Europe: an overview. Sustainability 5(6):2589–2608

Melillo JM, Gurgel AC, Kicklighter DW, Reilly JM, Cronin TW, Felzer BS, Paltsev S, Schlosser CA, Sokolov AP, Wan X (2009) Unintended environmental consequences of a global biofuels program. Report No 168. globalchange.mit.edu/files/document/MITJPSPGC_Rpt168.pdf. Accessed 10 Oct 2016

National Research Council (2009) A New Biology for the 21st Century. National Academies Press, Washington

O'Brien M, Schütz H, Bringezu S (2015) The land footprint of the EU bioeconomy: Monitoring tools, gaps and needs. Land Use Policy 47:235–246

OECD (2009) The bioeconomy to 2030 designing a policy agenda: main findings and policy conclusions. OECD International Futures Project. www.oecd.org/futures/long-termtechnologicalsocietalchallenges/42837897.pdf. Accessed 10 Oct 2016

OECD (2014) Working party on biotechnology issues paper for the OECD workshop on sustainable biomass drives the next bioeconomy. Directorate for Science, Technology and Industry Committee for Scientific and Technological Policy, OECD Organisation for Economic Co-operation and Development, Paris. www.oecd.org/officialdocuments/publicdisplaydocumentpdf/?cote=DSTI/STP/BIO(2014)3&docLanguage=En. Accessed 15 Oct 2016

Paula L, Birrer F (2006) Including public perspectives in industrial biotechnology and the biobased economy. J Agric Environ Ethics 19:253–267

Paustian K., Babcock, BA., Hatfield J., Lal R, McCarl, BA, McLaughlin S, Mosier A, Rice C, Robertson GP, Rosenberg, NJ, Rosenzweig C, Schlesinger WH, Zilberman D (2004) Agricultural mitigation of greenhouse gases: science and policy options. CAST (Council on Agricultural Science and Technology) Report. https://vtechworks.lib.vt.edu/bitstream/handle/10919/65238/153_agMitigaGreenGases.pdf?sequence=1 . Accessed 10 Oct 2016

Pavanan KC, Roeland AB, Cornelissen R, Philp JC (2013) Biomass sustainability and certification. Trends Biotechnol 31(7):385–387

Perella, M. (2016) 10 things you need to know about the circular economy. Guardian Sustainable Business. www.theguardian.com/sustainable-business/10-things-need-to-know-circular-economy. Accessed 10 Oct 2016

Petersen A, Krisjansen I (2015) Assembling 'the bioeconomy': exploiting the power of the promissory life sciences. J Sociol 51(1):28–46

Pfau SF, Hagens JE, Dankbaar B, Smits AJM (2014) Visions of sustainability in bioeconomy research. Sustainability 6:1222–1249

Ros J, Olivier J, Notenboom J, Croezen H, Bergsma G (2012) Sustainability of biomass in a bio-based economy. A quick-scan analysis of the biomass demand of a bio-based economy in 2030 compared to the sustainable supply. PBL Netherlands Environmental Assessment Agency. www.pbl.nl/sites/default/files/cms/publicaties/PBL-2012-Sustainability-of-biomass-in-a-BBE-500143001_0.pdf. Accessed 15 Oct 2016

Sauvé S, Bernard S, Sloan P (2016) The environmental sciences, sustainable development and circular economy: alternative concepts for transdisciplinary research. Environ Dev 17:48–56

Scarlat N, Dallemand J-F, Monforti-Ferario F, Nita V (2015) The role of biomass and bioenergy in a future bioeconomy: policies and facts. Environ Dev 15:3–34

Staffas L, Gustavsson M, McCormick K (2013) Strategies and policies for the bioeconomy and bio-based economy: an analysis of official national approaches. Sustainability 5(6):2751–2769

Schmid O, Padel S, Levidow L (2012) The bio-economy concept and knowledge base in a public goods and farmer perspective. Bio-based Appl Econ 1(1):47–63

Searchinger T, Heimlich R, Houghton RA, Dong F, Elobeid A, Fabiosa J, Tokgoz S, Hayes D, Yu TH (2008) Use of U. S. croplands for biofuels increases greenhouse gases through emissions from land-use change. Science 319(5867):1238–1240

Smith P, Martino D, Cai Z, Gwary D, Janzen H, Kumar P, McCarl B, Ogle S, O'Mara F, Rice C, Scholes B, Sirotenko O (2007) Agriculture. In Metz B, Davidson OR, Bosch PR, Dave R, Meyer LA (eds) Climate Change 2007: mitigation. Contribution of Working Group III to the fourth assessment report of the intergovernmental panel on climate change, Cambridge University Press, Cambridge/New York

Soetaert W, Vandamme E (2006) The impact of industrial biotechnology. Biotechnol J 1:756–769
The White House (2012) National bioeconomy blueprint. White House, Washington
United Nations (1992) AGENDA 21: United Nations conference on environment & development. Rio de Janeiro, Brazil, https://sustainabledevelopment.un.org/content/documents/Agenda21.pdf. Accessed 10 Oct 2016
von Braun J (2013) Bioeconomy – science and technology policy for agricultural development and food security. ppafest.nutrition.cornell.edu/authors/vonBraundraft.pdf. Accessed 10 Oct 2016

Part IV
The “Anthropocene” and Sustainable Development

Sustainable Development: Renaissance or Sunset Boulevard?

Massimiliano Montini and Francesca Volpe

1 Introduction

The sustainable development concept reached centre stage in the late twentieth and early twenty-first century, in large part due to the 1987 Brundtland Report and the subsequent 1992 Rio Conference. In the latter context, the concept was heavily emphasised within Agenda 21, the UN programme of action for the shaping of sustainable development patterns in the twenty-first century. Since Rio in 1992, the principle of sustainable development has played a pivotal role in the development and application of international and national policies; policies which are at the crossroads between economic development, social development and environmental protection. Notwithstanding this, sustainable development has gradually lost its propulsive impetus through the years. However, this situation could change soon, because of three independent, yet concatenate events which occurred in 2015. Those events could exercise a relevant influence on its future shaping. As a consequence, the year 2015 may ultimately be remembered as the turning point in the evolution of the concept of sustainable development.

The three above mentioned events consist of (a) the publication of Pope Francis's Encyclical Letter Laudato Sì—On care for our common home, (b) the adoption by the UN General Assembly of the Sustainable Development Goals (SDGs) and the related 2030 Agenda for Sustainable Development and (c) the conclusion of the Paris Agreement on Climate Change by the Conference of the Parties to the Framework Convention on Climate Change. In particular, the Encyclical Letter Laudato Sì brings back to centre stage the inherent ethical dimension of sustainable development, which has remained quite underdeveloped so far. The

M. Montini (✉) • F. Volpe
Department of Business and Law, Regulation for Sustainability Research Group (R4S), University of Siena, Siena, Italy
e-mail: massimiliano.montini@gmail.com; francescavolpe@gmail.com

L. Westra et al. (eds.), *The Role of Integrity in the Governance of the Commons*,
DOI 10.1007/978-3-319-54392-5_11

SDGs promote the pivotal role of sustainable development and dictate the global agenda for the promotion of sustainable development patterns for the next 15 years. The Paris Agreement on Climate Change recognises the need to integrate climate change and sustainable development considerations.

This chapter analyses the role that sustainable development plays in the three above mentioned documents and explores sustainable development's possible future evolutionary path. The next few decades will be crucial to determine the success or failure of the concept of sustainable development. On the basis of the analysis undertaken, this chapter presents a reference framework for addressing a fundamental question that needs urgent attention. Such a question may be expressed as follows: has the time finally come for the renaissance of the concept of sustainable development and the embracing of its meaningful application, after many years of uncertainty about its effective role and the lack of clarity as to the concept's meaning? Alternatively has the hey-day of the sustainable development principle passed? Like Norma Desmond in *Sunset Boulevard*, is it merely a dream that the sustainable development principle will enjoy a resurgence?

2 The Encyclical Letter Laudato Sì: On Care for Our Common Home

The Encyclical Letter "Laudato Sì. On Care For Our Common Home", issued by Pope Francis in May 2015, is a particularly timely document, which extends beyond purely religious relevance by addressing the present socio-ecological crisis (Laudato Sì 2015). Although not focused directly on the concept of sustainable development, Laudato Sì contains a robust critique of the current development model based as it is on a technocratic paradigm and dominated by purely economic considerations, irrespective of the environmental consequences. To this effect, Pope Francis highlights the need to redefine our notion of progress, since "*a technological and economic development which does not leave in its wake a better world and an integrally higher quality of life cannot be considered progress*" (para. 194). Moreover, the Encyclical Letter questions the technocratic paradigm as an "*undifferentiated and one-dimensional paradigm*" (para. 106) taken up by humanity, which lays the foundations for the overexploitation of natural resources, legitimising a potentially limitless economic growth. As acknowledged in the Encyclical Letter, "*the idea of infinite or unlimited growth, which proves so attractive to economists, financiers and experts in technology*", is misleading, since it is "*based on the lie that there is an infinite supply of the earth's goods*" (para. 106).

In such a context, Pope Francis takes a clear stance against the belief that "*current economics and technology will solve all environmental problems*" (para. 109), thus echoing the literature that warns against the myth of infinite growth in a planet characterised by limited resources and sinks for waste (Capra and Luisi 2014; Daly 1996a; Georgescu-Roegen 1986; Daly 1977; Schumacher 1973). Such literature is mainly grounded in the findings of the branch of physics known as thermodynamics. Its second law, the so-called entropy law, states that the availability of energy to perform useful work for human purposes is progressively reduced every time it passes through transformations, such as the ones brought about by economic activities (Ayres 1998; Daly 1996b; Faber et al. 1987; Georgescu-Roegen 1986; Georgescu-Roegen 1971). The combined understanding of the second law of thermodynamics along with the limited capacity of natural sinks to absorb waste as well as the finite nature of environmental resources shows the biophysical impossibility of limitless growth on a limited planet.

However, the mainstream economic model still fails properly to take into account its inherent contrast with the biosphere, the "*safe-operating space*", characterised by limited natural resources and limited sinks for waste and pollution (Rockström et al. 2009a, b; Tiezzi 2005; Daly 1996a; Costanza and Daly 1992). As Pope Francis clearly states, "*if in some cases sustainable development were to involve new forms of growth, then in other cases, given the insatiable and irresponsible growth produced over many decades, we need also to think of containing growth by setting some reasonable limits and even retracing our steps before it is too late*" (para. 193). Indeed, Pope Francis notes that, sometimes, notwithstanding economic growth, the quality of people's life may diminish due to the "*deterioration of the environment, the low quality of food or the depletion of resources*"; in such a case, "*talk of sustainable growth usually becomes a way of distracting attention and offering excuses*" (para. 194).

As acknowledged by Pope Francis, the global financial crisis of 2008 provided a tremendous opportunity for discontinuing such an economic model based on growth. However, its potential was not fully exploited and "*the response to the crisis did not include rethinking the out-dated criteria which continue to rule the world*" (para. 189). Therefore, Pope Francis's proposal to tackle the present crisis of overconsumption and exponential economic growth and reverse the current trend, is a "*renewal of humanity itself*" (para. 118): a renewal which is to be attained through the adoption of "*an integral ecology*" (para. 137). Such a concept is said to be composed of an environmental, economic, social, cultural ecology and daily life ecology (para. 137). However, the precise meaning of such an ecology "*remains somewhat elusive*" (Castillo 2016, p. 354). In any case, it can be seen as involving a call for a renewed relationship between humanity and nature. In fact, as acknowledged by the Encyclical Letter in its inception, "*nothing in this world is indifferent to us*". Therefore, humanity should seriously start reconsidering its relationship with the ecosystems to which it is a part. In such a context, sustainable development is mentioned as a fundamental and indispensable goal to be pursued within the required quest for the advocated change. To this effect, Pope Francis emphasises that "*the urgent challenge to protect our common home includes a concern to bring the whole human family together to seek a sustainable and integral development, for we know that things can change*" (para. 13).

In order to address such an urgent challenge, Pope Francis proposes some "*major paths of dialogue*", engaging politics, economy, religions and science, so as to escape the current "*spiral of self-destruction*" (para. 163). However, in Pope Francis's view, politics and laws will not be sufficient to stop the progressive deterioration of the environment unless we tackle human and social degradation. In fact, not only "*many things have to change course, but it is we human beings above all who need to change*". This highlights the necessity to deal with the "*great cultural, spiritual and educational challenge*" which "*stands before us*", by developing "*new convictions, attitudes and forms of life*" (para. 202). Pope Francis's wording is clear and advocates a "*profound interior conversion*" of humanity in the form of an "*ecological conversion*" (para. 217). The concept of ecological conversion is shaped throughout the Encyclical Letter by the spirituality of Saint Francis of Assisi (paras. 1, 10, 11, 12, 66, 87, 91, 125, 218, 221), who is defined by Pope Francis as "*the example par excellence* [...] *of an integral ecology lived out joyfully and authentically*" (para. 10). The advocated ecological conversion has to take place both at individual and societal levels; human beings are urged to rediscover "*the capacity to be happy with little*" and to embrace sobriety and humility (para. 222) with a view to create a new alliance between humanity and the environment which should replace the nowadays dominant "*alliance between the economy and technology*" (para. 54).

3 The Sustainable Development Goals and the UN 2030 Agenda for Sustainable Development

In September 2015, after a complex negotiating process started with the Rio + 20 Conference, the United Nations General Assembly adopted the Sustainable Development Goals (SDGs) and the related 2030 Agenda for Sustainable Development (SDGs 2015). The 17 SDGs, which are complemented by 169 targets, represent a "*comprehensive, far-reaching and people-centred set of universal and transformative*" objectives (SDGs 2015, para. 2). They build on the legacy of the Millennium Development Goals (MDGs), which aimed at providing a reference framework for action mostly focused on the eradication of poverty in the period 2000–2015. However, the SDGs substantially depart from the more limited approach which had been taken in the MDGs, since they address both developing and developed countries. In fact, all countries of the international community are called to define and implement their own national plans for sustainable development. The paramount reference to the concept of sustainable development is the other fundamental element that differentiates the SDGs from the previous MDGs. In fact, while the MDGs promoted development *tout-court*, the SDGs promote a revitalised global partnership for sustainable development as a common and shared goal of the international community. This is confirmed by the wording of the preamble of the 2030 Agenda for sustainable development, which shows a great determination of the international community "*to take the bold and transformative steps which are urgently needed to shift the world on to a sustainable and resilient path*" (SDGs 2015, Preamble).

Notwithstanding this general call for sustainable development addressed to all countries, the SDGs acknowledge and recall the importance of the principle of common but differentiated responsibilities (Principle 7 of the Rio Declaration) and promote the eradication of poverty as a fundamental objective. Within such a context, the 2030 Agenda characterises itself as a plan of action for people, planet and prosperity, based on the traditional three pillars structure of sustainable development. This is confirmed, once again, by the wording of the preamble of the Agenda, which contains a resolved commitment of the Parties "*to achieving sustainable development in its three dimensions – economic, social and environmental – in a balanced and integrated manner*" (SDGs 2015, Preamble).

The SDGs and the 2030 Agenda embrace the common understanding of sustainable development based on the three interlinked and integrated dimensions, namely the economic, social and environmental one. However, it is worth highlighting that an explicit definition of the sustainable development concept upon which the UN 2030 Agenda is grounded is lacking. Such lacuna represents a missed opportunity, insofar as within the SDGs the core of the sustainable development concept is maintained under a sort of a veil, preventing the interpreter from determining the proper relationship among the three dimensions of sustainable development. Moreover, the UN 2030 Agenda does not provide a reference point for the future work that has to be undertaken mostly at a national level for its implementation. Therefore, how the international community will be able to promote the advocated shift without any clear and commonly agreed definition of sustainable development remains an open question (Montini and Volpe 2015).

Moreover, it should be noted that despite the fundamental reference to sustainable development contained in its Preamble, the 2030 Agenda seems to be mainly concerned with the pursuit of (economic) growth as its paramount objective. In the preamble of the Agenda, the Parties "*envisage a world in which every country enjoys sustained, inclusive and sustainable economic growth and decent work for all*" (para. 9). Thus, the SDGs and the related 2030 Agenda contribute to maintaining the dangerous overlapping between the two concepts of development and growth. Indeed, it could be argued that they intentionally avoid clarifying the distinction between the "quantitative" nature of growth, which should be defined as "*an increase in through-put, which is the flow of natural resources from the environment, through the economy, and back to the environment as waste*", and the "qualitative" nature of development, which should be considered as "*a qualitative change, realisation of potential, evolution toward an improve but not larger structure or system – an increase in the quality of goods and services (where quality is measured by the ability to increase human well-being) provided by a given throughput*" (Daly and Farley 2011, p. 6; Costanza and Daly 1992). In this way they can keep on advocating the promotion of sustainable development whilst at the same time fostering the mainstream growth paradigm.

Any meaningful policy or set of actions promoting sustainable development should be grounded on a clear distinction between the opposed concepts of growth and development. Their assimilation gives rise to a certain confusion and runs counter to the well-established scientific finding that sustained economic growth can hardly be sustainable in a planet characterised by limited natural resources and sinks for waste. This suggests that the time has come for replacement of the misleading "growth oriented" pattern with an alternative "development oriented" pattern. Accordingly, the "*growthmania*" (Daly 1996a, p. 33) which has to date characterised the mainstream economic model should be abandoned. Unfortunately, the SDGs and the related UN 2030 Agenda seem substantially to ignore the impossibility of an unlimited growth on a limited planet (Daly and Farley 2011; Daly 1996a). As a consequence, we may argue that the SDGs, despite their notable positive features, are still embedded in a very traditional "growth economy" reference scenario, which may prove unsuitable in confronting the sustainable development challenge that the UN 2030 Agenda is called to address. Notwithstanding this, since the SDGs represent the commonly agreed plan for the promotion of sustainable development at the global level, they are in the position to provide a renewed impetus to the promotion of sustainable development in the next few years.

4 The Paris Agreement

The Paris Agreement on Climate Change, adopted in December 2015 and entered into force on 4 November 2016, represents a decisive departure from the previous approach to climate change, which was embedded in the 1997 Kyoto Protocol (Montini 2015). The Paris Agreement demonstrates a completely new attitude to the relationship between climate change and sustainable development. A close connection between climate change and sustainable development is established and the need to integrate climate change considerations into the sustainable development patterns of action is stressed. More specifically, the link between climate change and sustainable development is emphasised in the Preamble of the Paris Agreement, which acknowledges "*the intrinsic relationship that climate change actions, responses and impacts have with equitable access to sustainable development and eradication of poverty*". Moreover, article 2 of the Paris Agreement, when setting the context for its implementation, shows the explicit aim "*to strengthen the global response to the threat of climate change, in the context of sustainable development and efforts to eradicate poverty*". Furthermore, in article 4 of the Paris Agreement the Parties agree that the peaking of GHGs emissions and their subsequent reduction should be undertaken, *inter alia*, in the context of sustainable development. In addition to that, with specific regard to the connection between the Paris Agreement and the SDGs, it should be highlighted that Decision 1/CP.21 of the UNFCCC Conference of the Parties, which contains the adoption of the Paris Agreement, explicitly welcomes "*the adoption of United Nations General*

Assembly resolution A/RES/70/1, "Transforming our world: the 2030 Agenda for Sustainable Development", in particular its goal 13".

The close link between the climate change action goals and the objective of sustainable development is confirmed and reinforced also by article 6 of the Paris Agreement, which deals with the voluntary cooperation patterns that the Parties may develop in the framework of the implementation of the Agreement, in order to promote various form of joint fulfilment of their nationally determined contributions (NDCs). Within such a framework, the Paris Agreement introduces a new economic instrument, namely the "mechanism for sustainable development". Such a new mechanism incorporates sustainable development as one of its constitutive elements and primary goal. This way, it differentiates itself from the previous flexibility mechanisms under the Kyoto Protocol. In fact, the two project-based instruments featured by the Kyoto Protocol, namely Joint Implementation and Clean Development Mechanism, despite recognising the need to promote sustainability as one of the guiding elements that ought to underpin the realisation of projects, never really managed truly to contribute to the promotion of sustainable development (Voigt 2008, pp. 82–84). In this sense, it seems that the new mechanism, which incorporates sustainable development as a direct and primary goal related to the implementation of the projects, may be a decisive step forward, in so far as it represents great opportunity for a meaningful promotion of sustainable development in the climate change context.

More specifically the implementation of the new mechanism for sustainable development could pave the way for the promotion of a series of international investments resulting in a "double dividend". The double dividend may consist of a contribution to the mitigation of GHGs emissions, as well as leading to the promotion of sustainable development patterns in the countries where the related projects will be performed. Thus, the actions undertaken in the framework of the new mechanism could be a great opportunity for integrating climate change objectives and sustainable development goals as requested both by the Paris Agreements and the SDGs. In this sense, a paramount role is played by Goal 13 of the SDGs, which calls the international community to "*take urgent action to combat climate change and its impacts*" and in particular by the related Indicator 13.2, which advocates the need to "*integrate climate change measures into national policies, strategies and planning*". It is within such a context, that the implementation of Goal 13 of the SDGs, which will take place alongside the implementation of the Paris Agreement, might create a great opportunity to realise a positive intertwined advancement of both climate change and sustainable development objectives.

A further issue that should be discussed here is the tendency to couple the promotion of sustainable development with the call for economic growth, already examined above with regard to the SDGs, but which also is relevant to the Paris Agreement. In spite of its widespread recognition of the pivotal role of sustainable development and of its reference to environmental integrity contained in article 6, the Paris Agreement also incorporates a reference to the promotion of economic growth (art. 10). This again demonstrates the similar inconsistent approach already seen in the SDGs, which purportedly promote sustainable development and protect environmental integrity while, simultaneously remaining embedded within an economic growth paradigm. In this sense, we may conclude that the Paris Agreement, although to a lesser extent, resembles the approach of the UN 2030 Agenda, which does not dare to depart from the mainstream economic growth model.

5 Conclusion: Renaissance or Sunset Boulevard for Sustainable Development?

The occurrence of the three independent yet concatenated events which took place in 2015, may in the future be recognised as the turning point in the evolution of sustainable development concept (Capra 1987, p. 16).[1] Such a concept is a cornerstone of all the three analysed documents, namely the Encyclical Letter Laudato Sì, the SDGs and the related UN 2030 Agenda, and the Paris Agreement. This chapter has analysed them under the common thread of their contribution to sustainable development in order to set a reference framework for addressing the fundamental issue regarding the future of sustainable development.

In this sense, it may be held that the role that sustainable development is deemed to play in the near future will not depend merely on the independent legacy of the three events described above, but rather on their systemic integration and alignment.[2] In order to tackle the increasingly complex challenges posed by the evolution of the sustainable development concept, it will not be sufficient to act on the basis of a piecemeal methodology. Instead, a new approach based on the interplay of two opposing tendencies, an "integrative" tendency and a "self-assertive" one, should be pursued. An integrative tendency conceives of every element as part of the whole and takes into account the emerging properties that arise (only) from the interaction among the various parts. Meanwhile a "self-assertive tendency" values the unique features of each element (Capra and Luisi 2014, p. 65). In this way, a combined reading and understanding of the international agendas on climate change and on sustainable development should be promoted. Furthermore, there is a biunivocal relationship between the implementation of the respective targets set by the Paris Agreement and the SDGs. In fact, as argued in the literature, the implementation of many of the climate actions foreseen under the Paris Agreement would contribute to the achievement of specific SDGs targets, as well as the implementation of several SDGs targets would boost the possibility of countries

[1] On the basis of Capra's analysis, it may be said that a turning point can emerge when different elements are provided with a coherent conceptual framework which highlights their common features and aims.

[2] A call for a systemic integration and alignment of climate change and sustainable development in the implementation of both the UN 2030 Agenda and the Paris Agreement was made by representatives of governments, civil society and academia at a side event during the 2016 session of the UN High-level Political Forum on Sustainable Development, at http://sd.iisd.org/news/governments-stakeholders-discuss-alignment-of-sdgs-and-paris-agreement/.

to achieve the climate actions set out in the NDCs (Northrop et al. 2016, p. 13). In addition to that, the SDGs and the Paris Agreement should be read and implemented in conjunction with the Encyclical Letter. The latter could sustain both of them with its ethical afflatus and ground-breaking vision, while helping to fill their gaps and solving their inconsistencies.

The three events described above provide a new analytical framework against which the question regarding the future of the sustainable development concept should be addressed. In this sense, we may recall the fundamental question that we posed at the beginning of our analysis, namely whether the time has finally come for the renaissance of the concept of sustainable development or whether, quite on the contrary, it is destined in the medium-long term to the sunset boulevard. With regard to such a question, the next few years will be crucial for the concept's evolution and meaningful implementation.

In particular, only if the ethical, economic and environmental related dimensions of sustainable development are pursued in a comprehensive, integrated and trans-disciplinary way, will it be possible for sustainable development to flourish (Jørgensen et al. 2015). Should this happen, sustainable development might finally become a truly "grown-up concept" and complete its renaissance. If, conversely, such an integrated and trans-disciplinary evolution of the concept fails to emerge, it is very likely that sustainable development, despite its present popularity will retreat into 'the sunset boulevard'.

References

Ayres R (1998) Eco-thermodynamics: economics and the second law. Ecol Econ 26(2):189–209

Capra F (1987) The turning point. Bantam Books (originally Simon and Schuster, 1982)

Capra F, Luisi PL (2014) The systems view of life. A unifying vision. Cambridge University Press, Cambridge

Castillo DP (2016) Integral ecology as a liberationist concept. Theol Stud 77(2):353–376

Conference of the Parties to the UNFCCC, Decision 1/CP.21, Adoption of the Paris Agreement, 2015

Costanza R, Daly HE (1992) Natural capital and sustainable development. Conserv Biol 6 (1):37–46

Daly HE (1977) Steady-state economics. The economics of biophysical equilibrium and moral growth. W. H. Freeman and Company, San Francisco

Daly HE (1996a) Beyond growth. The economics of sustainable development. Beacon Press, Boston

Daly HE (1996b) Georgescu-Roegen versus Solow: Stiglitz, presentation at ISEE conference, Boston, August 1996

Daly HE, Farley J (2011) Ecological economics. principles and applications, 2nd edn. Island Press, Washington

Faber M, Niemes H, Stephan G (1987) Entropy, environment and resources. Springer-Verlag, Berlin

Georgescu-Roegen N (1971) The entropy law and the economic process. Harvard University Press, Cambridge

Georgescu-Roegen N (1986) The entropy law and the economic process in retrospect. East Econ J 12(1):3–25
Jørgensen SE, Fath BD, Nors Nielsen S, Pulselli FM, Fiscus DA, Bastianoni S (2015) Flourishing within limits to growth: following nature's way. Earthscan, London
Laudato Sì (2015) Pope Francis, Encyclical Letter Laudato Sì. On care for our common home
Montini M (2015) The Paris Agreement on climate change: miracle or disaster? Environ Liabil 23:161–166
Montini M, Volpe F (2015) Sustainable development goals: much ado about nothing?. Environ Liabil (4):141–147
Northrop E, Biru H, Lima S, Bouye M, Song R (2016) Examining the alignment between the intended nationally determined contributions and sustainable development goals. Working paper. World Resources Institute, Washington
Rockström J, Steffen W, Noone K, Persson A, Chapin FS III, Lambin E, Lenton TM, Scheffer M, Folke C, Schellnhuber HJ, Nykvist B, de Wit CA, Hughes T, van der Leeuw S, Rodhe H, Sörlin S, Snyder PK, Costanza R, Svedin U, Falkenmark M, Karlberg L, Corell RW, Fabry VJ, Hansen J, Walker B, Liverman D, Richardson K, Crutzen P, Foley J (2009a) Planetary boundaries: exploring the safe operating space for humanity. Ecol Soc 14(2):32
Rockström J, Steffen W, Noone K, Persson A, Chapin FS III, Lambin E, Lenton TM, Scheffer M, Folke C, Schellnhuber HJ, Nykvist B, de Wit CA, Hughes T, van der Leeuw S, Rodhe H, Sörlin S, Snyder PK, Costanza R, Svedin U, Falkenmark M, Karlberg L, Corell RW, Fabry VJ, Hansen J, Walker B, Liverman D, Richardson K, Crutzen P, Foley J (2009b) A safe operating space for humanity. Nature 461(7263):472–475
Schumacher EF (1973) Small is beautiful. A study of economics as if people mattered. Blond & Briggs, London
SDGs (2015) United Nations General Assembly (UNGA), Transforming our world: the 2030 agenda for sustainable development, resolution. Adopted on 25 September 2015
Tiezzi E (2005) Tempi storici, tempi biologici. Donzelli editore, Pisa
Voigt C (2008) Is the clean development mechanism sustainable? Some critical aspects. In: Sustainable development law & policy, Winter 2008, pp 15–21, 82–84

The Ecological Catastrophe: The Political-Economic Caste as the Origin and Cause of Environmental Destruction and the Pre-Announced Democratic Disaster

Donato Bergandi

WAR IS PEACE
FREEDOM IS SLAVERY
IGNORANCE IS STRENGTH
- George Orwell - *Nineteen Eighty-Four*
ENVIRONMENTALISM IS DEVELOPMENT[1]
- The Quintessence of the Sustainable Development Worldview

1 The Roots of the Ecological Disaster

We are living in a very advanced stage of a dystopian ecological catastrophe. At the planetary level, relatively few humans are consistently conscious of the deterioration of the life-support systems that allow life to persist and evolve on this planet.

In the meantime, the destruction of nature means enrichment for a tiny part of humanity, and goes hand in hand with severe environmental and economic losses for the great majority. This is particularly true in developing countries, where there still is an environment able to sustain higher levels of biodiversity in comparison to the environmentally more fragmented, economically advanced countries.

In any case, as a general trend, the populations of the developing and advanced countries must endure the exploitation of the planet's ecosystems and resources by an influential minority. In many cases, the populations have to face environmental disruption brought about by economic activities that are a source of enrichment for the few but the cause of pollution and environmental destruction for the many. Yet

[1]Another slightly more explicit way to express the attitude of the sustainable development paradigm towards nature is the following: PROTECTION OF NATURE IS DEVELOPMENT.

D. Bergandi (✉)
Muséum national d'Histoire naturelle, Paris, France
e-mail: donato.bergandi@mnhn.fr

L. Westra et al. (eds.), *The Role of Integrity in the Governance of the Commons*,
DOI 10.1007/978-3-319-54392-5_12

these same populations also contribute to the degradation of the environment because of their non-sustainable activities aimed at their survival and development.

It is becoming more and more evident, and even fastidiously trivial, that an influential minority, an oligarchic political-economic elite, a caste,[2] is in charge of all the economic flows and uses of biodiversity at the global and national level. Their management of resources is substantially autocratic and homogeneous, and above all it is, to a very great extent, independent of their declared political reference system of values. In other words, the political components of the caste may well belong to opposite poles of the political spectrum. Nevertheless, the economic strategies towards biodiversity are fundamentally the same and are generally free from any concern about the balance and integrity of ecological systems and the environmental and economic well-being of the general human population.

The cutting edge of this global environmental destruction is the ethical-political model of sustainable development; its undeclared purpose is to show that a new phase of human economic and cultural development has started, that "everything has changed", without genuinely calling into question the economic production system, which is the first and main cause of the ecological disaster we are living.

The sustainable development paradigm did not start with the Brundtland report (*Our Common Future,* 1987), but results instead from a long history that finds its roots in the nineteenth century and that gradually emerged from a succession of events that studded the twentieth century (Bergandi and Blandin 2012).[3] Gifford Pinchot (1865–1946), an American forester and politician, started the movement for the conservation of that nation's resources. His conception of conservation was essentially based on economic considerations. Conservation, for Pinchot, clearly and emphatically stands for "development". It is the affirmation of the right of any generation, present or future, to use the Earth's resources for their benefit. The actual model of sustainable development, with its recurrent reference to the conservation of resources and to future generations, has been clearly not only anticipated, but built, by Pinchot, at least in its normative basis: "Conservation does mean provision for the future, but it means also and first of all recognition of the right of the present generation to the fullest necessary use of all the resources that this

[2]This is a group of individuals made by tacit alliance between professional political elites and economic elites (industrialists, financial groups) who are fully aware of their interests, rights and privileges. Their social status is definable in terms of a new hybrid sociological space merging the categories of "corporation" and "caste". The latter is hereditary, while the hereditary dimension in the corporation is not necessarily present. The national and international political-economic oligarchy, with its sectarianism towards anyone not belonging to the group, has now acquired the pre-eminent features of a caste. The ideology and caste feeling imply a consciousness of belonging to a group with privileges and duties in a hierarchical context characterized by hereditary specialization and repulsion with regard to those who do not share this same social status (Bouglé [1935] 1969, pp. 3, 29).

[3]Other key moments in the history of the sustainability paradigm are the Biosphere Conference (1968), the Stockholm Conference (1972), the Rio de Janeiro Earth Summit (1992), the Johannesburg Summit (2002), and the Rio + 20 Earth Summit (2012).

country is so abundantly blessed with. It means the welfare of this generation and afterwards the welfare of the generations to follow" (Pinchot 1910, p. 42).

Conservation was also the prevention of waste: "So we are coming [...] to understand that the prevention of waste in all other directions is a simple matter of good business. The human race controls the earth it lives upon" (*Ibidem*, p. 45). Finally, the "public good" was the meta-principle to which all others were subordinated. Pinchot expresses his utilitarian credo along the lines of Jeremy Bentham (1748–1832) and John Stuart Mill (1806–1873): "Conservation means the greatest good to the greatest number for the longest time" (*Ibidem*, p. 48). It is interesting to note that, in the name of equality of opportunity for citizens, Pinchot also proffered a programmatic fight against the monopolies on the grounds that their influence could orient legislation in favor of their interests and against the public good (*Ibidem*, pp. 24–30).

But Pinchot's formal battle against the monopolies failed to stand the test of politics. In fact, following the devastating 1906 earthquake in San Francisco, the city needed water and electricity. Even though different possibilities existed for supplying these, Pinchot supported the economic groups who sought to build a dam in the Hetch Hetchy Valley in California's Yosemite National Park. John Muir (1838–1914), and all those who would protect the integrity of Yosemite Park, fought a media and political battle, but in the end "ravaging commercialism" (Muir 1912, p. 262) won the day, and the dam was built. Pinchot's conservationism-oriented management of nature and his priority on human use—"wise use", of course—of the planet's natural resources started down the road of the harmful, damaging ideological utopia that today we call "sustainable development".

2 Caste, Democracy and Nature

Before analyzing the significance and the practical impact of the sustainable development paradigm, in a propaedeutical way we must ask whether some ways of organizing power are more sensitive or intrinsically dangerous than others with respect to the protection of nature. Assuming that a totalitarian regime would represent the worst-case scenario, because a very limited number of people could decide on the environmental well-being of all the people, we limit the analysis to representative and direct democracy.

The harmful effects of the economic and financial activities of the lobbies and monopolies on nature are the ultimate consequence of a political danger that is always present in all forms of government. This danger has been very well identified by some of the founders of political liberalism. Bentham, treating the moral and political actions of individuals and communities, introduced a specific interpretative category to stigmatize the lethal danger of a full democracy, that of "sinister interests". These rogue interests privilege individual or group interests instead of encouraging the greatest happiness for the social community as a whole.

They tend to steer action in the wrong direction, in the sense that they determine unethical acts, acts of improbity that affect other people and are contrary to the interests of the community (Bentham 1815, §V, pp. 110–111). For his part, Mill, like Bentham, is extremely clear about the fact that no form of government is immune from the antisocial behaviour of elites in charge of the political management of the nation, and that such behaviour could harm and go against the development of the community: "One of the greatest dangers, therefore, of democracy, as of all other forms of government, lies in the sinister interest of the holders of power: it is the danger of class legislation, of government intended for (whether really effecting it or not) the immediate benefit of the dominant class, to the lasting detriment of the whole. [...] If we consider as a class, politically speaking, any number of persons who have the same sinister interest [...] the desirable object would be that no class, and no combination of classes likely to combine, shall be able to exercise a preponderant influence in the government" (Mill 1865, pp. 127–128). Along the same line of thought, in *On Liberty*, where he traces the history of the emergence of representative democracy in Europe, Mill develops the positive side of a legitimate ruling class: "What was now wanted was, that the rulers should be identified with the people; that their interest and will should be the interest and will of the nation" (Mill [1859] 1864, p. 10).

The lucid analysis by Bentham and Mill about the dangers that threaten representative democracy finds insightful echoes in the political philosophy and sociological analysis of Vilfredo Pareto (1848–1923), Gaetano Mosca (1858–1941), and Charles Wright Mills (1916–1962). First, it is philosophically and politically sounder not to use an interpretive filter of conspiracy theory, because the various actors belonging to a ruling class will, simply keeping their minds on their business and "following the lines of least resistance", find a convergent interest that shapes their political action: "That accord comes about automatically; for if in a given set of circumstances there is one line of procedure where the advantage is greatest and the resistance least, the majority of those who are looking for it will find it, and though each of them will be following it on his own account, it will seem, without being so, that they are all acting in common accord" (Pareto [1916] 1935, §2254, p. 1576).

Second, Mosca, for his part, considers that in political affairs the dominion of organized minorities over the unorganized majority is an inevitable outcome (Mosca 1896, chap. II, §4, p. 53). However, even if in our democratic societies the political struggle between individuals and groups is no longer a struggle for existence but a "struggle for pre-eminence" (Mosca 1896, chap. I, §13, p. 29; chap. V, §2, p. 121), we must keep in mind that when the governing class is the exclusive interpreter of the "will of the people" and no other organized social force can resist their "natural tendency" to abuse their powers, this means that "a despotism based ostensibly on popular sovereignty" has taken place (Mosca 1896, §6, p. 134).

Third, Mills, with his theory of the elite, sheds light on the sociological significance of institutional and social hierarchies in the United States and, more generally, in modern societies. His aim is "[...] to understand the facts of power and the ways of the powerful". The history of our political institutions results neither from a "blind drift" nor from a "conspiracy" of an unknown group of individuals (Mills [1956] 2000, p. 27). Alongside ordinary men and women who pursue projects that are beyond their control, the large majority, there is another category of men and women who rule the big corporations, who run the machinery of state, and who direct the military establishment. This "power elite" can make decisions that have major consequences, because they occupy the strategic command posts of the social structure (*Ibidem*, p. 4). The economic, the political and the military domains shape modern life via an "interlocking directorate" of people who are not always aware of their effective power (*Ibidem*, p. 8). However, at the top of the three enlarged and centralized domains, the economic, political and military elites form a compact social and psychological entity; they have become self-conscious members of a social class that tend to think and behave in very similar ways (*Ibidem*, p. 11). All this concentration of power in the hands of a few makes a mockery of democracy (Wolfe 2000, "Afterword", 364; similarly, John Stuart Mill considered that "exclusive government by a class" usurped the name of democracy, 1865, p. 163).

Political confrontation over the possession of a territory and its resources or for the pre-eminence of a specific political model once took the form of a gladiatorial fight, not unlike "nature red in tooth and claw". Today, although this is still a possibility, and unfortunately all too often an actuality, confrontation tends to be ritualized, and parliamentary debate takes the place of fighting. In a representative or indirect democracy, the exercise of legislative power is delegated to representatives of citizens who govern in the supposed interest of the people. If formally sovereignty lies in the people (i.e., in each citizen voter), in practice the exercise of the vote, the expression of freedom and autonomy, implies the delegation of decision-making power to representatives who should govern while pursuing the "common good". Paradoxically, then, in a system of representative democracy, the power of citizens is limited to delegating their "sovereignty" to representatives. But as Jean-Jacques Rousseau clearly noticed: "Sovereignty cannot be represented for the same reason that it cannot be alienated; it consists essentially in the general will, and the will cannot be represented; it is itself or it is something else; there is no middle ground. [...] The English nation thinks that it is free, but is greatly mistaken, for it is so only during the election of members of Parliament; as soon as they are elected, it is enslaved and counts for nothing" (Rousseau [1762] 2002, p. 221). *De facto*, the citizens, still formally sovereign at the moment of voting, lose their decision-making power to representatives, who are required to engage in governance that will set aside individual or group interests, and will focus, at least in theory, on the interests of the whole community.

In contrast, in the case of direct democracy, "government of all the people" means that all the people govern directly on their own: they participate directly in affairs of state without any mediation or intermediary between them and the law, without any alienation of the decision-making power of the people. The citizens are involved directly in lawmaking via petitions or referenda that they themselves initiate: they draft laws, place them directly on the ballot, and then reject or approve them (Haskell 2001, p. 49). A government by assembly exercises its functions without delegates, and any delegates that do exist have received specific, revocable, binding mandates from the electorate (Bobbio 1989, pp. 154, 155).

In the paradigmatic example of Athenian direct democracy, the officials were chosen by lot, they could serve "only for one year and no more than twice in a lifetime" (Osborne 2010, p. 28), and even if, as was very likely, not all citizens could participate in the Assembly, decisions were reached after debate (Thomas 2000, pp. 61–62). There were no professional politicians, because quick turnover maintained the system in a permanent state of healthy and reinvigorating, incorruptible "amateurism".

Even according to the normative terms of representative democracy, Western democracies resemble pseudo-democracies more than they do fully functioning democracies. A new political ethos able to involve the population in governing a democracy more directly should orient the political system towards incorporating specific directly democratic procedures. Certainly, one norm with decisive importance for a successful democratic adventure is the choice of "temporary" officials who would be "chosen by lot" a limited number of times in their lifetime (no more than once or twice). This rule prevents democracy from becoming a partitocracy, i.e., a system of governance by parties that ultimately leads to a focus on clan relationships and on the establishment of a parallel and masked underground power. The emergence of such an elite of political professionals is the very negation of democracy. The noteworthy reduction of the number of representatives in parliament should go hand in hand with a predominant recourse to direct initiatives, popular referenda and recalls (citizens can remove and replace public officials before the end of their term of office). Finally, officials would no longer receive their mandate from a political party, but directly from the electorate: a mandate that is binding and that can be revoked if not respected.

The current system of representative democracy is completely disconnected from what should be its main purpose, i.e., the pursuit of the "common good". The antidemocratic involution of the system is such that, for example, both the United States and Europe (the European Commission) have begun a "lobbying register" of the corporates that want to influence law-making! In this kind of context, sovereignty has ceased to be "popular" and has become instead "corporate sovereignty". Of course, the economic power of these corporates enables them to exercise disproportionate pressure on the decision-makers, obtain privileged access to information and steer European and US laws and rules. History tells us that the influence of the corporates on lawmakers greatly outweighs that of citizen associations. This system clearly supports private, "sinister" corporate interests against the social interests and right of the people to enjoy a healthy environment. To list just a few environmental issues where this holds true: pesticide use, genetically modified organisms, rock hydraulic fracking and pollution by the chemical industry.

In the context of representative democracy, the *sinister interests* of the ruling class, of the oligarchic political-economic caste, constituted by professional politicians and corporate firms, represent a danger for the balance of human society and for its harmonious development with respect to social justice and equality of opportunity. In a very similar way, at the level of the biosphere the impact of the human species on the biodiversity and life-support system is so extensive and devastating that, at least at first glance, we can talk about *human sinister interests* capable of altering evolutionary processes and the ecological balance of biotic communities. But in reality the political and economic choices of an influential minority are connoting the sense of action of the entire human species. An influential minority that presume that they are still fully within their rights even as they stubbornly continue to manipulate the political rules and violate natural laws.

3 Protection of Nature Is Development

The battle to build an international organization for the protection of nature started in 1913 and lasted 35 years.[4] After two world wars, the International Union for the Protection of Nature (IUPN) was finally founded at Fontainebleau, France, in 1948. But whatever hopes this organization might have represented for its preservationist supporters were short-lived. Unfortunately for the fate of the integrity of nature, at the IUPN general assembly in 1956, under pressure from the Anglo-Americans, the organization changed its name to the International Union for the Conservation of Nature and Natural Resources (IUCN). For those gathered, the "Protection of Nature" was sentimental and unrealistic (Holgate 1999, pp. 63–65).

In reality, this change of name is not so harmless as it might seem at first glance. It is emblematic of a deeper influential trend: Pinchot's "wise" conservationism won the political battle for the "management" of nature. With conservationism, which does not question at all the destructive reach of the dominant economic model, the roots of ecological catastrophe were planted. All this was happening in a very noxious way, because conservationism, and today its direct filiation, sustainable development, represent a political alibi that screens the true essence of things: they support in the present everyday reality the destruction of the very ecological systems that they are supposed to defend, in the name of a never realized, or still in progress, "wise use" or "sustainability".

Day after day, the dystopian ecological catastrophe is taking an increasingly Orwellian form. The mainstream way of thinking conveyed by the sustainable development paradigm proposes ways of living and policy making that are supposed to definitively solve the planetary problems of the erosion of biodiversity and climate change. What is the solution to all our environmental problems? In a very concise way the philosophy of sustainable development can be efficaciously summarized as follows: "Protection of nature is development". "Sustainable" development, of course.... This is a paradoxical and oxymoronic sentence, but, at first glance, why not? This is the inner, deep message of the sustainable development

[4]The Berne Conference for the International Protection of Nature in 1913 was the first international meeting on this issue.

"Newspeak" paradigm, carrying what George Orwell named "*Doublethink*" (Orwell's italicization), that is, "the power of holding two contradictory beliefs in one's mind simultaneously, and accepting both of them" (*Nineteen Eighty-Four,* [1949] 1959, p. 220). In fact, sustainable development policy implies *simultaneously* antithetic and irreconcilable theoretical positions: (1) the preservation of nature *and* the conservation of nature, that is, the sustainable developmental "wise use" of nature; (2) the recognition of the intrinsic value of nature, which involves the substantial intangibility and respect for the integrity of ecological systems *and* for the extrinsic value of nature (i.e., the use of nature as a function of human instrumental needs and aims) (see Preamble CBD, 1992); and (3) the maintenance of ecological sustainability, i.e., the capacity of ecosystems to retain their essential functions and processes and to maintain their biodiversity and evolutionary potential over the long term (resilience), *and* upholding economical sustainability, so-called sustainable developmental growth in a context of human demographic increase and a world of finite natural resources (Bergandi 2013).

The sustainable development paradigm is structured around this "doublethink" and what we can identify as its practical corollary, the "double-act".

From the theoretical point of view and the "doublethink" side, there is in fact a deep ethical tension and antinomy between the preservation of nature, the preservation of its evolutionary potential, the safeguard of the balance of ecological systems and support for all this at the same time that one is holding the idea that "nature" is *in essentiam* "natural resources". In other words, the complete philosophical and political failure of the sustainable development paradigm results from the deep "doublethink" inconsistencies that put an anthropocentric perspective side by side with the idea that ecological systems are bearers of values that transcend economic utilitarianism. Independently of any reference to categories such as "sustainability" or "intrinsic value", the underlying philosophical cornerstone of sustainable development is "anthropocentric", because the only happiness or well-being considered is, at best, that of all individuals of the human species, and "resourcist" because it considers that nature is *nothing but* a pool of resources and means that are functional to human well-being.

From the practical point of view and the "double-act" side, things become, if possible, even more complicated, because in the sustainable development Newspeak paradigm appears a word that in recent decades had almost vanished because it conveyed antinomian meanings with respect to ecological sustainability: the dreaded word is "growth"—but, of course, "growth that benefits all" (UNCSD Rio + 20 2012, p. 2). Other magic words that make their first appearance in this type of text include the "green economy", which "in the context of sustainable development [...] should contribute to meeting key goals [such as] sustained, inclusive and equitable growth [...] (*Ibidem*, p. 25).

Later, at the Conference of the Parties (COP 21) in Paris in 2015, after almost 50 years of international political exchanges on the approach to be adopted at the planetary level by environmental policy, the majority of the international community reached a consensus about the fact that "climate change represents an urgent and potentially irreversible threat to human societies and the planet and thus requires the widest possible cooperation by all countries" (FCCC/CP/2015/L.9). The international community agreed on the necessity "to strengthen the global response to the threat of climate change, in the context of sustainable development and efforts to eradicate poverty... to pursue efforts to limit the temperature increase to 1.5 °C above preindustrial levels" (Art. 2, FCCC/CP/2015/L.9).

There is clearly a discrepancy between the facts and the civilized, magnificent, almost transcendent world of international treaties. Following to the letter the paradigm of sustainable development in everyday life, we can readily ascertain that the satisfaction of the basic needs of the entire human population is still hypothetical, equality of access to natural and financial resources is not guaranteed, and the economy is doped because of distortions due to advantages gained by transnational economic groups at the planetary level. Above all, we can see clearly that the productivist, consumerist model of development, the very cause of the ecological disaster that we are living, continues to phagocytize the planet.

The recent past is not a harbinger of a reasonably optimistic future. In 2002, COP8 (New Delhi) made a commitment to achieve a significant reduction in the current rate of biodiversity loss at the local and global level by 2010. But in 2010, the 3rd Global Biodiversity Report found, for example, that the total abundance of vertebrates living in the tropics declined by 59% between 1970 and 2006. Some current clues tell us clearly that the ecological catastrophe is not only established and unequivocal, but that it is even gathering pace. Since the early 1990s, anthropogenic pressure has led to the loss of 10% of our planetary wilderness. Even though the area benefiting from global protection has expanded in the last two decades (by 2.5 million sq. km), an even greater area of wilderness has been lost (3.3 million sq. km). Some 27 ecoregions have lost any remaining significant wilderness area, particularly in South America, which experienced a 29.6% loss (Amazon basin), and Africa, with a 14% loss (e.g., Northwestern Congolian Lowland Forests) (Watson et al. 2016, pp. 1–2). Considering that half of all clearing in tropical forests between 2000 and 2012 was illegal, conservation interventions in wilderness areas need to be funded proactively to sustain the ecological integrity of these areas (*Ibidem*, p. 5; Westra et al. 2008).

Given mankind's efficiency at destroying planetary biodiversity, through the exploitation of species and the increasing fragmentation of the environment and through the impact of industrial activities on the planet's environmental life support systems, it is becoming increasingly clear that abandoning the resourcist anthropocentric ethics underlying our nature management policies—i.e. sustainable development and the Pinchotian conservationism of resources—is a moral, economic and political obligation.

It is a political obligation, because the resilience of what we can now name socio-ecosystems is no longer ensured by a representative system of popular sovereignty that disconnects, more and more, the environmental well-being of the populations and the preservation of the balance of natural systems from the decisions taken by professional politicians, who are oriented instead toward the preservation of their privileges and the interests of an influential minority.

It is an economic obligation, because, in the long term, the current approach based on the exploitation of resources, using techniques that are not conceived in function of minimizing the impact on ecosystems, is harmful not only to the biosphere, but also to the socio-economic systems themselves, because the only target sought is the benefit that can be derived in the short to medium term.

And finally, it is a moral obligation, because our value systems guide our actions in the world. If we want to build a new and more harmonious relationship with our environment, it is necessary to choose an ethical option in which humanity is not separate from the rest of nature, as many of our myths and religions have taught us for centuries. An ethical option would sustain the idea that "humans are nature" and, on the basis that there is a substantial ontological and evolutionary continuity between humans and nature, would construct a co-evolutive ecological ethics of "solidarity". In fact, humans are on this planet because of a concatenation of biotic and abiotic entities, processes and events. If our ethical options break this chain of solidarity to affirm a vaunted moral primacy of mankind, enshrining our erratic power over the rest of nature, our fate will, very likely, be sealed.

The current catastrophic conditions of the environment should be lived by humanity as a kind of moral challenge, a moral test emerging from our very recent and unruly technological and economic development. Our ability to meet this test will determine whether we are still able to live and evolve in our ecological niche, the planet Earth. We can solve the test positively only if our moral principles are in correspondence with the natural rules that govern the planet's ecological processes, our life and the life of our animal and plant companions on the evolutionary journey.

References

Bergandi D (2013) Epilogue: the epistemic and practical circle in an evolutionary, ecologically sustainable society. In: Bergandi D (ed) The structural links between ecology, evolution and ethics. The virtuous epistemic circle. Springer, Dordrecht, pp 151–158

Bergandi D, Blandin P (2012) De la protection de la nature au développement durable: genèse d'un oxymore éthique et politique (From the protection of nature to sustainable development: the genesis of an ethical and political oxymoron). Revue d'histoire des sciences 65(1):103–142. doi:10.3917/rhs.651.0103

Bobbio N (1989) Democracy and Dictatorship. The Nature and Limits of State Power. University of Minnesota Press, Minneapolis

Bouglé C (1969) Essais sur le régime des castes. Presses Universitaires de France, Paris

Bentham J (1815) A table of the springs of action. In: Goldworth A (ed) Deontology; together with a table of the springs of action; and the article on utilitarianism. Clarendon Press/Oxford University Press, New York/Oxford

Haskell J (2001) Direct democracy or representative government? Westview Press, Boulder

Holgate M (1999) The green web: a union for world conservation. Earthscan Publications, London

Mill JS (1864) On liberty, 3rd edn. Longman, Green, Longman, Roberts & Green, London

Mill JS (1865) Considerations on representative government, 3rd edn. Longman, Green, Longman, Roberts & Green, London

Mills CW (2000) The power elite. Oxford University Press, Oxford/New York

Mosca G (1896) The ruling class. McGraw-Hill Book Company, New York/London
Muir J (1912) The Yosemite. The Century Co, New York
Orwell G (1959) Nineteen eighty-four. Secker and Warburg, London
Osborne R (2010) Athens and Athenian democracy. Cambridge University Press, Cambridge
Pareto V (1935) The mind and society, vol 4, 1st edn. Harcourt, Brace and Company, New York
Pinchot G (1910) The fight for conservation. Doubleday, Page & Company, New York
Rousseau J-J ([1762] 2002) The social contract and the first and second discourses. Edited and with an Introduction by Susan Dunn; with essays by May, G., R.N. Bellah, D. Bromwich and C.C. O'Brien. Yale University Press, New Haven/London
Thomas R (2000) The classical city. In: Osborne R (ed) Classical Greece 500–323 BC. Oxford University Press, Oxford, pp 52–80
UN Conference on Environment and Development (1992) Convention on biological diversity, Rio de Janeiro, Brazil. Adopted 5 June 1992
UN Conference on Sustainable Development (Rio+20) (2012) The future we want. Resolution 66/288 adopted by the General Assembly on 27 July
UN Framework Convention on Climate Change (2015) 21st session of the conference of the parties to the UNFCCC (COP 21). Adoption of the Paris Agreement on 12 December 2015
Watson JE et al (2016) Catastrophic declines in wilderness areas undermine global environment targets. Curr Biol 26:1–6
Westra L, Bosselmann K, Westra R (2008) Reconciling human existence with ecological integrity. Earthscan, London
Wolfe A ([1956] 2000) Afterword. In: Mills CW (ed) The power elite. Oxford University Press, Oxford/New York, pp 363–381
World Commission on Environment and Development (1987) Our common future. Oxford University Press, Oxford/New York

Ecological Integrity in the Anthropocene: Lessons for Law from Ecological Restoration and Beyond

Geoffrey Garver

Earth's life systems face an expanding aggregation of anthropogenic impacts that by many measures has already overshot global ecological limits (Rockström et al. 2009; Carpenter and Bennett 2011; Berkhout 2014; Steffen et al. 2015). Humanity is caught between irreconcilable narratives of impossibility regarding the human prospect in the Anthropocene—a proposed geological epoch defined by human impacts (Crutzen and Stoermer 2000; Steffen et al. 2011). The currently dominant growth-insistent narrative says it is impossible to improve human and social welfare, and to protect the environment, unless the human economy grows indefinitely. The limits-insistent narrative says infinite economic growth inevitably leads to global ecological impacts catastrophic to present and future generations of human and other life, and that a downsizing and stabilization of the economy is urgently needed to reverse global ecological trends and restore degraded ecological functions. The Anthropocene prompts an urgent need to move towards a limits-insistent narrative that gives primacy to containing the human enterprise within ecological limits in a life-enhancing and mutually enhancing human-Earth relationship (Berry 1999)—whereby humans see themselves as members, not masters, of the entire community of life on Earth and interact with Earth and the life it supports respectfully and "for the benefit of the larger community as well as ourselves" (Berry 1999, p. 5).

The emerging concept of "safe operating space for humanity" within planetary boundaries (Rockström et al. 2009; Steffen et al. 2015) aligns with the limits-insistent narrative because it recognizes that global ecological boundaries, not insistence on economic growth, constrain socio-political and economic options.

G. Garver (✉)
School of Environment, McGill University, Montreal, QC, Canada

Department of Geography, Planning and Environment, Concordia University, Montreal, QC, Canada
e-mail: gginmont@videotron.ca

L. Westra et al. (eds.), *The Role of Integrity in the Governance of the Commons*,
DOI 10.1007/978-3-319-54392-5_13

Other concepts relevant to the notion of a mutually enhancing human-Earth relationship include ecological integrity, "right relationship" (Brown and Garver 2009) and "safe and just operating space for humanity" (Rockström et al. 2009; Raworth 2012; D'Alisa et al. 2015; Steffen et al. 2015). In law, the notion of a mutually enhancing human-Earth relationship has inspired the concepts of Earth Jurisprudence and Wild Law (Cullinan 2011), which scholars continue to elaborate and develop (Burdon 2011; Burdon 2015).

This chapter provides an analysis and critique of the notion of a mutually enhancing human-Earth relationship, ecological integrity and other terms noted above. These concepts all inevitably incorporate some degree of human impact because of the long history of human transformation of ecosystems. Their use in the development of innovative approaches to law and governance must recognize this fact.

1 A Mutually Enhancing Human-Earth Relationship as Humanity's Overarching Goal

A mutually enhancing human-Earth relationship reflects the notion that "in reality there is a single integral community of the Earth that includes all its component members whether human or other than human [each of which] has its own role to fulfill, its own dignity, its inner spontaneity" (Berry 1999, p. 4). It is a relationship in which "[e]very being enters into communion with other beings" (Berry 1999, p. 4). With regard to law, Berry put forth the overarching notion that "[e]cology is not part of the law; law is an extension of ecology" (Berry 1999, p. 84)—in other words, that law should reflect and maintain a human role within the broader community of life that is life-enhancing and respectful of the ecological role of other members of that community. Every component of the Earth community would have a right to exist (Cullinan 2011) and a right "for habitat and the opportunity. .. to fulfill its role in the natural system to which it belongs" (Berry 1999, p. 80). Humans and all other components of the Earth system would be mutually responsible to respect all other components, and "[j]ustice would consist in carrying out this complex of creative relationships" (Berry 1999, pp. 61–62).

Berry's proposal for a mutually enhancing human-Earth relationship lacks clear criteria for determining when the relationship is mutually enhancing, or for adjudicating conflicting rights claims among components of the Earth community. For example, without more clarity, one could argue that as long as they are present on Earth, even in the current era of human degradation of many interacting processes within the Earth system (Ellis and Ramankutty 2008; Rockström et al. 2009; Steffen et al. 2015), humans are "participating members of the comprehensive Earth community" (Berry 1999, p. 8)—albeit, at times not very nice ones. Because of the human potential both to foster and hinder a mutually-enhancing human-Earth relationship, more clarity is needed to guide individual and collective human intention and choice.

More refined notions of membership in the Earth community, such as notions of householding, membership and entropic thrift (Brown 2015), will help make this idea practicable. These notions reinforce the need for humans to see the Earth as a cherished home, to identify themselves as integral members in a community of life and to be cautious and careful in the use of material and energy needed to sustain them. As well, the proposal for rights of each member of the Earth community to fulfill its role in the functioning of natural systems invites a complex, multi-scale and cross-scale inquiry into what that role is, and what systems are "natural" at different scales.

2 Toward a Human-Inclusive Ecocentric Paradigm

Fostering a mutually enhancing human-Earth relationship is not yet a prevalent societal goal at any scale, as overarching commitments in major international texts to sustained economic growth make clear (United Nations 2012; United Nations 2015; G20 2015). The Anthropocene highlights the extent to which humans have impaired the capacity of humanity and Earth's ecosystems to be mutually enhancing and brings into sharp focus the need to move past the current dominant narrative of the human-Earth relationship and the human behaviors it promotes as good and worthwhile.

The dominant anthropocentric narrative gives primacy to economic growth and technological innovation, reinforced by legal systems built around strong notions of state sovereignty and private property rights (M'Gonigle and Takeda 2013; Burdon 2015; Capra and Mattei 2015). At the extreme, it is a narrative of humanity "'mastering' nature to reclaim Eden" (Merchant 2003, p. 3)—at least in Western worldviews that ground economics and other constructs in a biblically derived paradigm of orderly, Cartesian human dominion over nature (Merchant 1980; Ost 2003; Ponting 2007; Capra and Mattei 2015). This narrative is fatally flawed because ultimately, the construction of artificial nature destroys the capacity of the ecosphere to support the flourishing of human and other life. Unintended consequences ultimately will overwhelm the impossible quest to dominate, control and perfect nature thoroughly. The surprise and irreducible uncertainty inherent in the evolution of socio-ecological systems are too greatly ignored, and individual human preferences and technological optimism too greatly relied upon.

The Anthropocene also calls into question the purist ecocentric narrative that laments the "decline from a pristine earth to a paved, scorched, endangered world" (Merchant 2003, p. 2) and calls for the restoration of wild nature on Earth. This narrative idealizes a pristine world without human impacts—and hence, without humans. Such a "cult of wilderness" (Miller 2001) narrative is fatally flawed for two main reasons. First, it ignores the significant extent to which humans already have dominated and altered Earth's ecosystems. More than three quarters of Earth's ice-free land surface is made up of "anthropogenic biomes" in which humans have altered the structure and function of ecosystems through agriculture, forestry, urbanization and other interventions (Ellis and Ramankutty 2008; Alessa and Chapin 2008). Second, if applied to its logical extreme from the local to the global level, the cult of wilderness narrative leaves no meaningful place for humans in Earth's life systems.

A narrative based on a mutually enhancing human-Earth relationship must be situated between these two imperfect narratives. Human-inclusive ecocentric approaches will be needed to yield "[n]ew kinds of stories, new ways of thinking, and new ethics" (Merchant 2003, p. 4) for the Anthropocene, ones that promote socio-ecological integrity and a mutually enhancing human-Earth relationship. They must strive for a thriving human presence within a life-enhancing global ecosystem, while rejecting both ecocentrism that is indifferent to human life and hubristic anthropocentrism that ignores the complexity of the human-Earth relationship.

Merchant (2003) frames the human-inclusive ecocentric narrative in terms of a human-Earth partnership grounded in an ethic that "holds that the greatest good for the human and nonhuman communities is in their mutual living interdependence" (Merchant 2003, p. 223). This ethic extends moral standing, equity and rights of inclusion and participation to nonhuman communities and species, as well as to all human communities and cultures (Merchant 2003). It implies that a notion such as ecosystem services (to humans) must be complemented with the notion of human services to ecosystems. This notion of mutual service must be part of a mutually enhancing human-Earth relationship in the Anthropocene.

3 "Ecological Integrity" as a Core Objective

The concept of ecological integrity helps clarify broader concepts like sustainability or a mutually enhancing human-Earth relationship. Many international agreements or soft law instruments refer to ecological integrity as an overarching, or at least significant, objective (Kim and Bosselmann 2013). For example, the Earth Charter is grounded in part on a call to "protect and restore the integrity of Earth's ecological systems, with special concern for biological diversity and the natural processes that sustain life" (Earth Charter 2000). The Earth Charter broadly aligns ecological integrity with prevention of environmental harm using a precautionary approach; adoption of "patterns of production, consumption, and reproduction that safeguard Earth's regenerative capacities, human rights, and community well-being [; and] [a]dvance[ment of] the study of ecological sustainability and promot[ing] the open exchange and wide application of the knowledge acquired" (Earth Charter 2000). The outcome document at the Rio + 20 conference on sustainable development, *The Future We Want*, called for "holistic and integrated approaches to sustainable development which will guide humanity to live in harmony with nature and lead to efforts to restore the health and integrity of the Earth's ecosystem" (United Nations 2012).

Ecological integrity and related notions of integrity are also integral to other broad proposals for an overarching ethic for the Anthropocene. Leopold's enduring environmental ethic is grounded in enhancing the integrity of the biotic community (Leopold 1949). Brown and Garver (2009) reiterate this moral mandate to ensure "right relationship" within the Earth system—where right relationship refers to a "guidance system for functioning in harmony with scientific reality and enduring ethical traditions" (Brown and Garver 2009, p. 4). The normative frame of safe operating space for humanity that underlies planetary boundaries proposals includes consideration of biosphere integrity (Steffen et al. 2015). Kim and Bosselmann (2013) contend that, with appropriate refinement and clarity, ecological integrity could emerge as the supreme objective of law in the Anthropocene.

4 Accounting for Historical Human Impacts in Defining Ecological Integrity

Ecological integrity—and related concepts such as biological integrity (Angermeier and Karr 1994), biosphere integrity (Steffen et al. 2015), integrity of the biotic community (Leopold 1949), biome integrity (Mace et al. 2014) and ecosystem integrity (De Leo and Levin 1997)—tends to connote notions of health, wholeness and naturalness (Angermeier and Karr 1994; Westra et al. 2000; Ordonez and Duinker 2012; Bridgewater et al. 2015). Ecological or biological integrity is usually defined with reference to pristine (or relatively pristine) ecosystems, such as wilderness areas or other protected areas that are far from human settlements and relatively free of anthropogenic impacts (Ordonez and Duinker 2012)—for example, "wild nature ... that is virtually unchanged by human presence or activities" (Westra et al. 2000, p. 20). So defined, ecological integrity potentially can be assessed by comparing an ecosystem's structure and function with benchmark conditions in a comparable ecosystem that is considered relatively free of human impacts, or in the same ecosystem prior to disturbance, taking into account natural variability (Angermeier and Karr 1994; Westra et al. 2000; Ordonez and Duinker 2012).

The Anthropocene underscores the near certainty that ecosystems completely free of anthropogenic impacts no longer exist—and that widespread human impacts on "wild nature" have occurred since Paleolithic times (Ellis and Ramankutty 2008; Minteer 2012; Ellis et al. 2013; Foley et al. 2013; Ruddiman et al. 2014; Lewis and Maslin 2015). A fully operable definition of ecological integrity must take into account the full history of human impacts on ecosystems from the local to global scale. A properly nuanced and scalable understanding of ecological integrity must encompass not only the most pristine wild ecosystems, but also dense human settlements and other areas in which humans have significantly transformed the evolutionary trajectory as compared to an imagined human-free ecosystem (Ellis and Ramankutty 2008). Perhaps in recognition of the lack of stark delineations, benchmarks of ecological integrity have sometimes been tied to ecosystems that are "virtually" pristine or "relatively" free of anthropogenic impacts (Westra et al. 2000). By departing from a pure notion of integrity, these qualifiers highlight the need for additional criteria.

Ecological integrity must be assessed according to systems' long-term cycles and functions (Gunderson and Holling 2002; Brown and Garver 2009). As well, the spatial scale at which ecological integrity must be considered ranges from the global to the very local. The most pristine ecosystems are part of higher-scale landscapes and ecosystems, and ultimately the entire ecosphere (Westra et al. 2000). The anthropogenic impacts or alterations in those larger landscapes and ecosystems at some point become significant. For example, "[n]o completely evolutionary processes may persist in an urban area (although reproduction of some species continues), but successful work has been done to show that such areas *can* be made compatible with integrity" (Westra 2011, p. 332). Yet, the criteria for compatibility in such a case remain elusive. References to biophysical conditions of the Holocene reflecting global ecological integrity (Rockström et al. 2009), because they "preceded the anthropogenic global environmental change that began with the Industrial Revolution" (Kim and Bosselmann 2013, p. 306), point in the right direction but may still be insufficient in light of global anthropogenic impacts that occurred or began prior to the Industrial Revolution.

Temporal scale presents other complexities. Any practicable definition of a mutually enhancing human-Earth relationship must account for temporal trends within and across spatial scales in a vast, integrated complex of social and ecological systems—that is, the global panarchy (Gunderson and Holling 2002). Complex adaptive systems inevitably evolve, typically with cyclical changes in the arrangements of structure and function (Gunderson and Holling 2002). As well, the path dependence of systems limits the scope of possible futures given past systems conditions, and the past-future divide that defines path dependence is constantly advancing. The resulting irreducible uncertainty in the evolution of feedbacks and other systems dynamics across temporal scales, from the local to the global, makes it impossible to predict the performance of the human-Earth system over time. Thus, giving priority to an enduring human-Earth partnership requires a broad, coordinated and adaptive system of human norms from the local to the global level, taking into account temporal factors.

A rigorous yet practicable notion of ecological integrity must acknowledge and accept a conditional level of symbiosis between humans and non-human nature in order to be consistent with a mutually enhancing human-Earth relationship. Studies showing human impacts on Earth systems dating back to Neolithic or Paleolithic times (Ellis and Ramankutty 2008; Ellis and Trachtenberg 2013; Foley et al. 2013; Ruddiman 2013) suggest that for much of human history, at different times and in different places, such a relationship may well have existed—and therefore, that environmental history can help reveal factors that either help or impede it (Tainter 1990; McNeill 2003; Diamond 2005). The challenge becomes determining where, along the spectrum from the least impacted to the most anthropogenically transformed ecosystems, practicable benchmarks for ecological integrity should be drawn at different spatial and temporal scales, accounting for humans as an integral ecosystem component. This determination necessarily reflects the aggregate of the individual and collective intentions of human beings. Understanding the role of intention in successful ecological and eco-cultural restoration and in examples of sustainable governance of common pool resources (CPRs)[1] provides direction for a global approach for instilling a mutually enhancing human-Earth relationship.

5 Lessons from Ecological Restoration and Governance of Common Pool Resources

Experiences in ecological restoration and eco-cultural restoration comprise a rich source for translating broad ideals tied to ecological integrity into practicable approaches at various scales, taking into account historical information regarding the human-Earth relationship. Both ecological and eco-cultural restoration involve a "process of assisting the recovery of an ecosystem that has been degraded, damaged, or destroyed" (SER 2004), taking into account criteria such as ecological integrity, historical fidelity and community engagement (Higgs 2003). Compared to ecological restoration, eco-cultural restoration places more emphasis on recovery of mutually supportive cultural practices and ecosystem structure and functioning. Both ecological and eco-cultural restoration inherently emphasize that restoration efforts always require a communal choice as to the desired objectives and the means to achieve them.

Human intention is key. Inherent in the proactive domains of ecological and eco-cultural restoration is a human dimension that encapsulates not only the reality of historical human impacts on ecosystems from the local to the global level, but also the deliberate choice that is involved in pursuing "a vision of a better relationship between humans and the rest of the world" (Egan et al. 2011, p. 1). Because of the deliberate human choice involved, ecological or eco-cultural restoration "is inherently (1) value-laden, (2) context driven, (3) prone to be immersed in disagreement and compromise, and (4) experiential" (Egan et al. 2011, pp. 1–2).

Although restoration ecology at times links ecological integrity to pristine human-free nature (Higgs 2003), ecological integrity more commonly has a broader connotation that accounts for human presence in ecosystems. Higgs describes integrity as "an all-encompassing term for the various features—resilience, elasticity, stress response, and so on—that allow an ecosystem to adjust to environmental change" (Higgs 2003, p. 122). This notion of integrity inherently includes human-

[1]CPRs are resources (e.g. water, timber) to which access is restricted to a defined community, unlike open access resources, to which access is unrestricted.

induced environmental change, although with conditions and limits. A particularly elegant expression of the dialectic this entails is the notion of reciprocal restoration (Geist and Galatowitsch 1999; Egan et al. 2011). Reciprocal restoration is "the mutually reinforcing restoration of land and culture such that the repair of ecosystem services contributes to cultural revitalization and renewal of culture promotes restoration of ecological integrity" (Kimmerer 2011, p. 258).

Higgs (2003) accounts for a nuanced interpretation of ecological integrity in advancing an approach to ecological restoration grounded in the integrated concepts of focal practice, ecological integrity, historic fidelity and intention. He captures the conditional space for ecologically acceptable human impacts within the concept of ecological integrity by framing it broadly around these factors, while avoiding the malleable notion of ecological health—which is more detached from alignment with historical conditions and trajectories, and therefore more prone to subjective interpretations (Higgs 2003; Westra 2011; Goldberg et al. 2015). The commitment to re-establishing historical eco-cultural trajectories also distinguishes ecological restoration from ecological engineering or design, in which "novel ecosystems" (if they have not already emerged unintentionally) might be created without reference to historical conditions (Balaguer et al. 2014). Higgs (2003) embraces a flexible range of options that include a human presence that maintains (1) a basis for mutual flourishing of the human and non-human spheres of nature, (2) ties to historical conditions in which a mutually enhancing human-Earth relationship existed, (3) insistence on core notions of resilience and persistence in accordance with broad historical socio-ecological trajectories, and (4) recognition of embedded, resilient and communal connections of people and place.

This flexibility in the meaning of ecological integrity can be incorporated into the selection of reference ecosystems that are used to establish objectives for restoration, or more broadly for socio-ecological integrity. Thus, Balaguer et al. (2014) advocate for selection of "*locally-tailored* historical references, using all the available and appropriate conceptual tools, so as to integrate both latent and on-going ecological and socio cultural processes and values" (Balaguer et al. 2014, p. 13, emphasis in original). However, because ecological restoration is often concerned with re-establishing a damaged ecosystem's historical trajectory as it was prior to the human disturbance of concern, the human impacts that may be incorporated into an appropriate historical reference are limited to those that were integral to that pre-existing trajectory (Higgs 2003; Balaguer et al. 2014).

A discursive and expansive notion of ecological integrity resonates with studies of social systems that have maintained sustainable, enduring use of at least some CPRs in supporting ecosystems (Ostrom 1990). From these studies, Ostrom identified key—and essential—features of social systems that maintain human provisioning from ecosystems of at least some human needs in perpetuity, without reliance on privatization or government regulation. Those interdependent features (Ostrom 1990, p. 90) are:

- **"Clearly defined boundaries.** ... Individuals or households who have rights to withdraw resource units from the CPR must be clearly defined, as must the boundaries of the CPR itself."
- **"Congruence between appropriation and provision rules and local conditions.** ... Appropriation rules restricting time, place, technology, and/or quantity of resource units are related to local conditions and to provision rules requiring labor, material, and/or money."
- **"Collective-choice arrangements.** ... Most individuals affected by the operational rules can participate in modifying the operational rules."
- **"Monitoring.** ... Monitors, who actively audit CPR conditions and appropriator behavior, are accountable to the appropriators or are the appropriators."
- **"Graduated sanctions.** ... Appropriators who violate operational rules are likely to be assessed graduated sanctions (depending on the seriousness and context of the offense) by other appropriators, by officials accountable to these appropriators, or by both."
- **"Conflict-resolution mechanisms.** ... Appropriators and their officials have rapid access to low-cost local arenas to resolve conflicts among appropriators or between appropriators and officials."
- **"Minimal recognition of rights to organize.** ... The rights of appropriators to devise their own institutions are not challenged by external government authorities."
- **"Nested enterprises.** ... Appropriation, provision, monitoring, enforcement, conflict resolution, and governance activities are organized in multiple layers of nested enterprises."

Ostrom emphasized that the emergence of these features, and of the collective constraint on human demands on supportive ecosystems, is contingent on a strong communal attachment to place and the local community, and to intergenerational continuity. These place-based and intergenerational commitments imply a need to adapt legal and governance structures and rules for sustainable and shared human use of supportive ecosystems in response to information from monitoring of relevant social and ecological indicators at a local scale (Ostrom 1990).

A challenge in applying a cohesive yet practicable notion of ecological integrity is that ecosystems at the local, landscape and regional level inevitably and increasingly are subject to human impacts that derive from spatially and temporally diverse drivers—including those that are remote geographically or whose impacts are temporally delayed. Thus, at whatever temporal or spatial scale, practical applications of ecological integrity to promote a mutually enhancing human-Earth relationship require an adaptive, multi-scalar systems approach that maintains an ongoing focus on and connection to the local scale.

6 Conclusions

The analysis in this chapter leads to several key conclusions regarding the foundations of a mutually enhancing human-Earth relationship, including an operable understanding of ecological integrity. First, any practicable notion of ecological integrity must include space for humanity as an integral part of nature and the Earth's ecosystems. Second, achieving a mutually enhancing human-Earth relationship necessarily requires clear and coherent criteria for controlling individual and collective intentions and choices. Properly done, ecological restoration adheres to an adaptive, systems-based approach and both commits to and recognizes the need for human intention, choice, decision making and active intervention to restore ecosystems that human activity has damaged. Third, to be consistent with an ethic oriented toward achieving and maintaining a mutually enhancing human-Earth relationship, individual and collective human choices must be informed by contemporary science, and especially an understanding of the behavior of complex adaptive systems. This approach requires the humility that is inherent in dealing with the inevitable uncertainties and unpredictability in how systems evolve by applying a precautionary approach, and not the hubris of many Western secular and religious traditions that idealize total human mastery and perfection of nature.

References

Alessa L, Chapin S III (2008) Anthropogenic biomes: a key contribution to earth-system science. Trends Ecolo Evol 23(10):529–531

Angermeier PL, Karr JR (1994) Biological integrity versus biological diversity as policy directives. Bioscience 44:690–697

Balaguer L, Escudero A, Martin-Duque JF, Mola I, Aronson J (2014) The historical reference in restoration ecology: re-defining a cornerstone concept. Biol Conserv 176:12–20

Berkhout F (2014) Anthropocene futures. Anthropocene Rev 1(2):154–159

Berry T (1999) The great work: our way into the future. Three Rivers Press, New York

Bridgewater P, Kim RE, Bosselmann K (2015) Ecological integrity: a relevant concept for international environmental law in the Anthropocene? Yearb Int Environ Law 25(1):61–78

Brown PG (2015) Ethics for economics in the Anthropocene. In: Brown PG, Timmerman P (eds) Ecological economics for the Anthropocene. Columbia University Press, New York

Brown PG, Garver G (2009) Right relationship: building a whole earth economy. Berrett-Koehler, San Francisco

Burdon P (ed) (2011) Exploring wild law: the philosophy of earth jurisprudence. Wakefield Press, Kent Town

Burdon PD (2015) Earth jurisprudence: private property and the environment. Routledge, New York

Capra F, Mattei U (2015) The ecology of law: toward a legal system in tune with nature and community. Berrett Koehler, San Francisco

Carpenter SR, Bennett EM (2011) Reconsideration of the planetary boundary for phosphorus. Environ Res Lett 6:1–12

Crutzen PJ, Stoermer EF (2000) The Anthropocene. Glob Chang Newsl 41:17–18

Cullinan C (2011) Wild law: a manifesto for earth justice, 2nd edn. Chelsea Green, White River Junction VT

D'Alisa G, Demaria F, Kallis G (eds) (2015) Degrowth: a vocabulary for a new era. Routledge, New York

De Leo GA, Levin S (1997) The multifaceted aspects of ecosystem integrity. Conserv Ecol 1(1):3. http://www.consecol.org/vol1/iss1/art3/

Diamond J (2005) Collapse: how societies choose to fail or succeed. Viking Press, New York

Egan D, Hjerpe EE, Abrams J (eds) (2011) Human dimensions of ecological restoration: integrating science, nature, and culture. Island Press, Washington

Ellis EC, Ramankutty N (2008) Putting people in the map: anthropogenic biomes of the world. Front Ecol Environ 6(8):439–447

Ellis EC, Trachtenberg Z (2013) Which Anthropocene is it to be? Beyond geology to a moral and public discourse. Earth's Future 2:122–125

Ellis EC, Kaplan JO, Fuller DQ, Vavrus S, Goldwijk KK, Verburg PH (2013) Used planet: a global history. Proc Nat Acad Sci 110(20):7978–7985

Foley SF, Gronenborn D, Andreae MO, Kadereit JW, Esper J, Scholz D, Pöschl U, Jacob DE, Schöne BR, Schreg R, Vött A, Jordan D, Lelieveld J, Weller CG, Alt KW, Gaudzinski-Windheuser S, Bruhn K-C, Tost H, Sirocko F, Crutzen P (2013) The Palaeoanthropocene – the beginnings of anthropogenic environmental change. Anthropocene 3:83–88

G20 2015 G20 Leaders' Communiqué, Antalya Summit [online]. http://www.g20.utoronto.ca/2015/151116-communique.html

Geist C, Galatowitsch SM (1999) Reciprocal model for meeting ecological and human needs in restoration projects. Conserv Biol 13(5):970–979

Goldberg MS, Garver G, Mayo NE (2015) Revisiting the metaphor of human health for assessing ecological systems and its application to ecological economics. In: Brown PG, Timmerman P (eds) Ecological economics for the Anthropocene. Columbia University Press, New York

Gunderson LH, Holling CS (eds) (2002) Panarchy: understanding transformations in human and natural systems. Island Press, Washington

Higgs E (2003) Nature by design. The MIT Press, Cambridge

Kim RE, Bosselmann K (2013) International environmental law in the Anthropocene: towards a purposive system of multilateral environmental agreements. Transnat Environ Law 2:285–309

Kimmerer R (2011) Restoration and reciprocity: the contributions of traditional ecological knowledge. In: Egan D, Hjerpe EE, Abrams J (eds) Human dimensions of ecological restoration: integrating science, nature, and culture. Island Press, Washington

Leopold A (1949) A sand county Almanac. Oxford University Press, New York

Lewis SL, Maslin MA (2015) Defining the Anthropocene. Nature 519:171–180

M'Gonigle M, Takeda L (2013) The liberal limits of environmental law: a green legal critique. Pace Environ Law Rev 30(3):1005–1115

Mace GM, Reyers B, Alkemade R, Biggs R, Chapin FS III, Cornell SE, Diaz S, Jennings S, Leadley P, Mumby PJ, Purvis A, Scholes RJ, Seddon AWR, Solan M, Steffen W, Woodward G (2014) Approaches to defining a planetary boundary for biodiversity. Glob Environ Chang 28:289–297

McNeill JR (2003) Observations on the nature and culture of environmental history. Hist Theory 42(4):5–43

Merchant C (1980) The death of nature: women, ecology and the scientific revolution. Harper & Row, San Francisco

Merchant C (2003) Reinventing Eden: the fate of nature in western culture. Routledge, New York

Miller C (2001) An open field. Pac Hist Rev 70(1):69–76

Minteer BA (2012) Geoengineering and ecological ethics in the Anthropocene. Bioscience 62 (10):857–858

Ordonez C, Duinker PN (2012) Ecological integrity in urban forests. Urban Ecosyst 15:863–877

Ost F (2003) La nature hors la loi: L'écologie à l'épreuve du droit. La Découverte/Poche, Paris

Ostrom E (1990) Governing the commons: the evolution of institutions for collective action. Cambridge University Press, New York

Ponting C (2007) A new green history of the world: the environment and the collapse of great civilizations. Penguin Books, Toronto

Raworth K (2012) A safe and just space for humanity: can we live within the doughnut? Oxfam Discussion Paper [online]. https://www.oxfam.org/sites/www.oxfam.org/files/dp-a-safe-and-just-space-for-humanity-130212-en.pdf

Rockström J, Steffen W, Noone K, Persson Å, Chapin FS III, Lambin E, Lenton TM, Scheffer M, Folke C, Schellnhuber H, Nykvist B, De Wit CA, Hughes T, van der Leeuw S, Rodhe H, Sörlin S, Snyder PK, Costanza R, Svedin U, Falkenmark M, Karlberg L, Corell RW, Fabry VJ, Hansen J, Walker B, Liverman D, Richardson K, Crutzen P, Foley J (2009) Planetary boundaries: exploring the safe operating space for humanity. Ecol Soc 14(2):32. http://www.ecologyandsociety.org/vol14/iss2/art32/

Ruddiman WF (2013) The Anthropocene. Annu Rev Earth Planet Sci 41:45–68

Ruddiman R, Vavrus S, Kutzbach J, He F (2014) Does pre-industrial warming double the anthropogenic total? Anthropocene Rev 1(2):147–153

Society for Ecological Restoration International Science & Policy Working Group 2004 The SER international primer on ecological restoration. Society for Ecological Restoration International. Tucson. www.ser.org

Steffen W, Grinevald J, Crutzen P, McNeill J (2011) The Anthropocene: conceptual and historical perspectives. Phil Trans R Soc A 369:842–867

Steffen W, Richardson K, Rockström J, Cornell SE, Fetzer I, Bennett EM, Biggs R, Carpenter SR, de Vries W, de Wit CA, Folke C, Gerten D, Heinke J, Mace GM, Persson LM, Ramanathan V, Reyers B, Sörlin S (2015) Planetary boundaries: guiding human development on a changing planet. Science 347(6223):1259885-1-1259885-10

Tainter JA (1990) The collapse of complex societies. Cambridge University Press, Cambridge

United Nations (2012) The future we want. United Nations, New York. https://documents-dds-ny.un.org/doc/UNDOC/GEN/N11/476/10/PDF/N1147610.pdf?OpenElement

United Nations (2015) Transforming our world: the 2030 agenda for sustainable development. United Nations, New York

Westra L (2011) Governance for integrity? A necessary but distant goal. In: Burdon P (ed) Exploring wild law: the philosophy of earth jurisprudence. Wakefield Press, Kent Town, pp 324–336

Westra L, Miller P, Karr JR, Rees WE, Ulanowicz RE (2000) Ecological integrity and the aims of the global integrity project. In: Pimentel D, Westra L, Ness RF (eds) Ecological integrity: integrating environment, conservation, and health. Island Press, Washington, pp 19–41

Part V
Human Responsibility for Ethical Governance

Addressing the Problem of Conflict-of-Interest and Moneyed Influence in Public Health: Some Case Studies

Colin L. Soskolne

1 Introduction

The International Joint Policy Committee of the Societies of Epidemiology (hereinafter "IJPC-SE") was founded in Seattle at the 2nd North American Congress of Epidemiology in 2006. Its focus is on *health for all through ethical, independent and transparent science.*

Founded as a consortium of national and international professional societies/associations of epidemiology, the IJPC-SE works in the public interest at the interface of research and policy. It helps to inform the optimal translation of science into effective public policy. The IJPC-SE's mission includes promoting the ethical and unbiased application of epidemiology, striving to bring clarity to the science of epidemiology in order to pave the way for rational, evidence-based policy. We work to promote and protect public health by serving as an ethical and effective counterweight to the misuse of epidemiological evidence.

2 Conflict-of-Interest and Disclosure

Epidemiology is most critical for informing public health policy because it bridges findings from research using animal models to the human experience. As an applied "soft" science (as opposed to a basic "hard" science), the science of epidemiology can be misapplied, misused, misinterpreted and abused by those intent on so doing.

C.L. Soskolne (✉)
University of Alberta, Edmonton, AB, Canada

Health Research Institute, University of Canberra, Canberra, ACT, Australia
e-mail: colin.soskolne@ualberta.ca

L. Westra et al. (eds.), *The Role of Integrity in the Governance of the Commons*,
DOI 10.1007/978-3-319-54392-5_14

The process of peer review aims to correct honest error in keeping science consistent with its mission of pursuing truth. However, those with agendas different from the pursuit of truth, usually intent on fomenting uncertainty and casting doubt on the evidence, can manipulate epidemiology in ways that derail science meant to serve the public interest. Their purpose is usually to misguide those empowered to formulate policy, with their main intent being to delay policy in favor of maintaining the *status quo*. By maintaining the *status quo*, business interests are generally protected, at least in the short-medium-term.

So strong today is moneyed influence in the derailment of science that, while public policy lags, the public's health and the sustainability of life on Earth are in jeopardy; in jeopardy, because evidence to better protect health and the environment is marginalized in favor of the *status quo*. The importance, therefore, of the IJPC-SE—as well as equivalent such entities across a range of public interest organizations—cannot be over-emphasized. These entities are needed as counterweights to moneyed influence used to preserve powerful interests, usually corporate in nature.

This chapter brings to attention how moneyed influence is being used to infiltrate the epidemiological literature in order to derail the scientific discourse and its role in public policy formulation. The need for disclosure of real or perceived Conflict-of-Interest (COI) must thus be taken seriously especially if those engaged in public health are to protect the public interest.

In essence, honest disclosure, enforced through editors as gatekeepers of the legitimate scientific literature (as opposed to predatory journals) is essential if those using science are to be aware of the extra-care needed in their use of a class of knowledge called "junk science". Junk science is often published by predatory journals, or by journals whose care in peer review is less than optimal. In addition, the infiltration onto legitimate journal editorial boards and review panels of scientists who have been seduced by money and have rationalized their move away from science's mission to pursue truth, is a mechanism to be countered also through declarations of COI (Ruff 2015).

3 Conflict-of-Interest (COI)

Conflict-of-Interest, hereinafter abbreviated to COI, is also referred to as "conflicts of interest", "conflict of interests", or "conflicting interests". In the field of scientific knowledge and its use in informing public policy, the significance of COI relates to knowledge transfer which is the domain of those tasked with translating science into policy. In epidemiology, some refer to its application for informing policy as "consequential epidemiology".

Attempts to define these are, as with most ethical questions, context-related. In the context of publication ethics, the World Association of Medical Editors states that:

> COI exists when there is a divergence between an individual's private interests [competing interests] and his/her responsibilities to scientific and publishing activities such that a reasonable observer might wonder if the individual's behaviour or judgment was motivated by consideration of his/her competing interests.

The International Society for Environmental Epidemiology (ISEE) 2012 *Ethics Guidelines for Environmental Epidemiologists*[1] address the topic of COI from several angles:

> (a) Avoiding Conflicting Interests: A conflict-of-interest occurs whenever a political, bureaucratic, career, or economic incentive, real or perceived, has the potential for producing partiality or compromising objectivity ...
> (b) Full Disclosure: Environmental epidemiology researchers, Institutional Review Board (IRB)/Research Ethics Board (REB) members, journal reviewers and editors, research grant reviewers, and other professionals who have a decision-making or primary role in the funding, conduct, or publication of research should provide full disclosure of financial and/or other advisory relationships that could influence their decision-making ...
> (c) Guidelines Governing Disclosure: Institutions, funding agencies, regulatory agencies, journals and other organisations that control the conduct, publication, or implementation of environmental epidemiology research should support the goal of transparency by establishing guidelines governing disclosure of relationships that may underlie all types of conflicting interests.

In sum, a conflict exists whenever an epidemiologist's role, obligation, or personal interest in accommodating an institution, sponsor, job/financial security, or personal goals compromises obligations to others who have a right to expect objectivity and fairness. Such circumstances are to be scrupulously avoided in conducting environmental epidemiology investigations because the attendant negative health consequences in environmental epidemiology research can be great.

COI can arise in every avenue of the epidemiologist's endeavor, including:

(a) Serving as a peer reviewer
(b) Publishing
(c) Receiving financial support
(d) Election to office and having a vote
(e) Serving on boards and on review/editorial/advisory panels
(f) Interpreting the evidence before us
(g) Receiving donations (from vested interests)

For the ethical professional, there are certain watch words to which attention must be paid to avoid being accused of an undisclosed COI: that which constitutes *real* COI vs. *perceived* COI, and on the need for *disclosure* and/or *recusal*. Damage caused from failures to disclose or recuse oneself has ramifications at two levels: personal and public.

Personal harm resulting from a failure to disclose or recuse oneself when one ought to have done so, include:

(a) Loss of credibility

---

[1] http://www.iseepi.org/About/documents/ethics_guidelines_adopted_april_25_2012_001.pdf.

(b) Loss of job
(c) Banned from publishing or receiving funds for a specified time

Public harm results from:

(a) The infiltration of advisory/expert boards/panels by those with a COI, and
(b) A literature infiltrated with junk and poor science. The effect of this is seen when the literature, upon which policy decisions depend, is contaminated with junk science that succeeds in casting doubt on the body of knowledge, making policy recommendations to better protect the public's health less likely.

4 A Recent Case Study from Montreal, Quebec, Canada

Ruff publishes material on the ongoing challenge of addressing ethical and human rights issues relating to asbestos in Canada and abroad. In particular, at RightOn-Canada.ca[2] on July 25, 2015, Ruff archived correspondence on Concordia University in Montreal and its role in promoting the use of asbestos. This archived correspondence provides a rich case study of the improper influence of asbestos industry interests and the failure of the University to respond appropriately when the matter was brought to its attention.

Concordia University is one of two major English-speaking universities in Quebec, Canada (the other being McGill University). In 2010, a consortium of investors sought a \$58 million loan guarantee from the Quebec government to re-open a Quebec asbestos mine with a view to exporting asbestos to developing countries. The leader of the consortium, Baljit Chadha, and his wife, Roshi Chadha (both presidents of companies exporting asbestos from Quebec at that time) sat on the Board of Governors of Concordia and McGill universities.

Baljit Chadha made significant financial contributions in the form of donations to Concordia's John Molson School of Business. He was named Governor Emeritus, and had an auditorium named in his honor. His contributions to Concordia were reflected in this way.

A lecturer at Concordia's John Molson School of Business, John Aylen, who had his own public relations company, was hired by Chadha as a public relations consultant and spokesperson to promote Chadha's asbestos project. The Quebec Medical Association, all of Quebec's Directors of Public Health, and the scientific community opposed or found significant fault with Chadha's project.

In 2014, Concordia University commissioned, funded, published and praised a report which eulogized Chadha and his project to export asbestos. The report stated that the scientific evidence supports the use of chrysotile asbestos and those who opposed the project acted irrationally, unethically and emotionally. The report was written by John Aylen. There was no disclosure that he had been hired by Chadha as

[2] http://www.rightoncanada.ca/?p=3072.

a consultant to promote the asbestos mine. There was no disclosure that Chadha was a Governor Emeritus and major donor to Concordia University.

A formal complaint was made to Concordia University explicating the above-noted scientific and ethical improprieties. It was submitted to Concordia University, challenging the report's scientific misinformation, industry influence, and undisclosed Conflict-of-Interest. Concordia was asked to retract its report because it served to contribute to public harm. Concordia was asked to carry out an independent, transparent investigation as to how such serious violations of academic integrity had occurred in order to prevent future occurrences. Concordia refused to do so. Instead, Concordia carried out a secretive, internal pretense of an "investigation" whereby Concordia's School of Business "investigated" itself.

Such conduct is seen as an overall system failure of ethical oversight; it is deemed unethical by people who believe that science and ethics require openness and transparency. This case study is an example of *the fox guarding the henhouse*.

Concordia University's "investigation" condoned the wrongdoing, providing no information about the supposed "investigation"; its investigation was not made public. Further, no action was taken, apart from a vague, facile statement that Concordia had addressed the COI issue. No details were provided, and the report was not retracted. Instead, Concordia extolled its publication of harmful scientific misinformation as being "the essence of academic freedom".

To be complete, soon after the complaint was made to Concordia University, the report was taken down from the University's website. No explanation was provided as to why a report, which Concordia had been promoting as an example of excellent work and which had been presented at a national conference, had suddenly been removed. Concordia refused requests to retract the report. The response of the University was thus to simply try to "disappear" the offensive report, in essence hiding the fact that it was published by a supposedly reputable institution, re-writing history, and thus avoiding any accountability for the wrongdoing. The report, in fact, had been available on the Concordia University website to be downloaded by whoever had wanted to make use of it. It remains an official publication of the University, commissioned and funded by the University and published on University letterhead, and can still be cited as such.

At a systems level, academia is a multi-billion dollar industry that fails to adequately address COI. At the end of the day, ethics are key to science and to public health. Without a serious, effective system of ethical oversight, evidence and public policy are being easily corrupted by vested interests. Overwhelming evidence of this has been provided by, among others, the tobacco industry, the asbestos industry, the fossil fuel industry, the chemical industry, the fast food industry, the mining industry, the sweetened beverage industry, and the lead industry.

Another even more egregious case study of a systemic failure in ethical oversight—also relating to asbestos and emanating from Quebec, Canada—relates to McGill University and was published by Ruff (2014); it is more egregious because it has endured for decades.

5 The IJPC-SE's Approach to Disclosure for Protecting Against COI

All members of the Board are required to provide a duly-completed Disclosure Statement[3] which is then considered by the Executive Committee and the Board. If approved, it is posted to the IJPC-SE website.[4]

The Disclosure Statement was produced as an "interim form" to be replaced should a more suitable form be developed in the future. The form provides a definition of terms relating to COI. The key element of the form is that the proposed Board member signs to the effect that she/he will recuse her/himself should a COI exist. The key wording follows:

> If in my service to the IJPC-SE I should ever find myself in a conflict-of-interest, actual or perceived, I will declare it and offer to recuse myself from all circumstances that could in any way influence Board deliberations as per the Founding Bylaws (Article V, Section 2).

This makes contractual the expectation of full disclosure and recusal when warranted.

6 A Current IJPC-SE Conundrum: A Real or Perceived COI?

Early in 2016, the IJPC-SE secured 501(c)(3) status as a public charity in the USA. In launching its first fundraising campaign, it was necessary to produce a Gift Acceptance Policy[5] to make transparent those entities from whom donations would and would not be acceptable. Guided by its Gift Acceptance Policy, donated monies should not be able, or even interpreted to be used to influence the work of the IJPC-SE; nor should they be used to tarnish its reputation as an objective public interest organization pursuing truth for the application of evidence in knowledge translation.

The offer of a donation from a law firm that specializes in representing plaintiffs with mesothelioma in relation to asbestos exposure was made early in 2016. The law firm has made use, over the past 4 years, of the IJPC-SE's 2012 Position Statement on Asbestos[6] as supportive evidence in their litigation. Soon after the

---

[3]https://www.ijpc-se.org/documents/IJPC-SE_TEMPLATE_DISCLOSURE_STATEMENT_October_7,_2016.pdf.

[4]https://www.ijpc-se.org/?p=about.

[5]https://www.ijpc-se.org/documents/IJPC-SE_Financial_Contribution_Policy-June_8_2015_revision_approved_June_26_revised_Sep_16_2015.pdf.

[6]https://www.ijpc-se.org/documents/03.JPC-SE-Position_Statement_on_Asbestos-June_4_2012-Full_Statement_and_Appendix_A.pdf.

IJPC-SE had secured charity status, the law firm pledged to make a donation of 20,000 USD to the IJPC-SE.

The IJPC-SE Gift Acceptance Policy, in its current form, is silent on donations from law firms. The IJPC-SE Executive Committee was conflicted as to whether the IJPC-SE could accept such a donation without the appearance of impropriety or risk of accusations of conflict-of-interest; namely that

(1) there could be an actual or perceived conflict-of-interest;
(2) the IJPC-SE may be considered to be beholden to the vested interests of particular donors; and
(3) ultimately, that the objectivity of the IJPC-SE and the public health impact of its work products such as Position Statements—which represent its major field of activity and may be used in litigation—could be called into question.

Therefore, on July 20, 2016, the ICPC-SE Board charged the Development Committee with producing an Addendum to its existing Gift Acceptance Policy.

The mission of the IJPC-SE is to generate position statements and policy recommendations based on the objective evaluation and analysis of scientific evidence, representing an authoritative and consensus opinion. However, certain entities view independent, objective scientific evidence as challenging and potentially threatening to their vested interests, usually financial in nature. Such entities have endeavored to undermine the IJPC-SE by disparaging the mission, products, and individuals associated with its Board for advancing independent, objective scientific evidence and recommendations for action that challenge the harmful misinformation campaigns of their particular vested interest.

The courtroom is an adversarial setting in which financial stakes are high, and the policy implications are great. Attorneys will pursue any tactic possible to undermine, discredit, and challenge the admissibility of opinions, testimony, or evidence that they perceive as a threat to their positions. This would include Position Statements of the IJPC-SE. Past or present financial contributions, or support from individuals, organizations, corporations, or other entities are included in the targets that may be cited as evidence of bias on the part of the IJPC-SE.

The asbestos industry, in particular, has already and will likely continue to attempt to undermine the IJPC-SE because of the IJPC-SE 2012 Position Statement on Asbestos, which advances independent, objective scientific evidence that exposes and challenges the industry's misinformation campaigns and potentially threatens their vested interests. The IJPC-SE must be aware that such attacks are likely to be repeated as the IJPC-SE continues its mission to uphold independent, objective scientific evidence, and the public good.

It is recognized, however, that without a funding base, the IJPC-SE will not be able to sustain itself and fulfill its public interest mission. Thus, the IJPC-SE is committed to establishing an endowment such that the investment income will sustain it in perpetuity. All possible donations consistent with its Gift Acceptance Policy must therefore be considered. In the context of working toward the establishment of a large enough endowment (in the region of 3 million USD), donations from philanthropic foundations are the preferred source, and are expected to constitute the greatest portion of the anticipated amount. Individual, smaller donated amounts are necessary, and are considered as seed monies demonstrating a groundswell of professional, stakeholder, and general public support prior to the IJPC-SE making what then could be successful applications to philanthropic foundations. These individual-level donations are also considered to be an important continuing source of support and engagement with the IJPC-SE after foundation funding is secured.

In summary, a proposed ADDENDUM to the existing IJPC-SE Gift Acceptance Policy is currently being debated by the IJPC-SE's Development Committee. The pros and cons of accepting any offer of a donation from a law firm is being discussed. No decision has been reached at the time of this chapter going to press. The point of including a conundrum such as this is to share the challenges that being a charity can pose to a public interest scientific organization. The IJPC-SE expects to rise to the challenge through being clear and transparent both in its rationale and in its conduct.

7 Conclusion

Epidemiology is a public health science used for informing policy. Increased vigilance and oversight are needed, particularly in the face of moneyed influence in the conduct and application of the discipline. All that can be done to maximize the ethical and transparent use of epidemiology—in the face of financial influence and possible misuse of the science—is critical if the public interest is to be protected. Public funding to support such initiatives would be optimal for avoiding real or perceived conflicting interests. The mission of the International Joint Policy Committee of the Societies of Epidemiology (IJPC-SE) includes protecting the public interest by serving as a counterweight to the improper use of the discipline.

Acknowledgements While the IJPC-SE's Executive Committee, its Development Committee, and other colleagues have contributed to thinking about COI and Disclosure in terms of its Gift Acceptance Policy, the contributions of colleagues Shira Kramer, Kathryn Gwiazdon, Martin Strosberg, and Kathleen Ruff to the current version of the proposed Gift Acceptance Policy Addendum are especially noted with thanks.

Disclosure

- As a professional epidemiologist, I have served as an expert witness in litigation on behalf of plaintiffs in the past, monies from which generally went into a University-managed research account; and the text portion associated with each of these two bullets of DISCLOSURE should appear immediately to the right of the respective bullet, thus saving a couple of lines.
- As a professional legacy, I bankrolled the International Joint Policy Committee of the Societies of Epidemiology (IJPC-SE) as a voluntary professional organization from 2011–2016, in the hope that it might become a self-sustaining charity in perpetuity. This chapter describes an aspect of the work in support of the mission of the IJPC-SE.

Postscript

1. A text edited by Martin J. Walker is expected to be published early in 2017. It addresses in far more depth than seen in this chapter the same and related issues. It is titled: Corporate Ties That Bind: An Examination of Corporate Manipulation and Vested Interest in Public Health and will be published by Skyhorse Publishing, Inc., New York, New York (See: http://skyhorsepublishing.com/titles/11632-9781510711884-corporate-ties-that-bind).
2. In April 2017, a further complaint was submitted to Concordia University. The complaint related to the fact that its asbestos report was being quoted by the International Chrysotile Association (ICA) in its lobbying material at a United Nations meeting as part of the ICA's efforts to prevent the listing of chrysotile asbestos as a hazardous substance under the Rotterdam Convention. In response, Concordia put a notice on its website stating that the report had been withdrawn from the website because the author's conflict-of-interest had not been disclosed in the report. Concordia side-stepped its failure to uphold academic and ethical standards, or its failure to carry out an independent, transparent investigation. Concordia never retracted the report; it simply withdrew it from its website (See: http://www.concordia.ca/about/administration-governance/updateon-report-withdrawn-from-website.html).

References

Ruff K (2014) Asbestos: a continuing failure of ethics by McGill University. Int J Occup Environ Health 20(1):1–3. Editorial. http://www.tandfonline.com/doi/full/10.1179/1077352513Z.000000000102

Ruff K (2015) Scientific journals and conflict of interest disclosure: what progress been made? Environ Health 14:45. Commentary. https://ehjournal.biomedcentral.com/articles/10.1186/s12940-015-0035-6

Ethics and Pesticides: The Precautionary Principle as Illustrated by Glyphosate

Josef Unterweger

In the course of the approval process for glyphosate eight European environmental organisations—GLOBAL 2000 (Austria), Nature & Progrès Belgique (Belgium), Générations Futures (France), Pesticide Action Network UK, Pesticide Action Network Europe, Pesticide Action Network Germany, WeMove Europe, Umweltinstitut München (Germany)—filed a complaint against Monsanto, the German Federal Institute for Risk Assessment or BfR, and the EFSA, the European Food Safety Authority.

The following looks at the approval procedure for pesticides in the European Union, with the application for approval of glyphosate as an example.

The first part will cover glyphosate, the main pesticide manufacturers that produce it, and the European Union's approval procedures. The second will highlight the timeline of the approval procedure and the various steps taken by the environmental organisations during this procedure.

After that, there will be an insight into the reasons for the complaints and their consequences. This will be followed by a brief résumé.

Glyphosate[1] is known under a number of different brand names, but the most familiar is Roundup. Glyphosate is used all over the world as a weed killer in agriculture, horticulture, industry, and also by private households. It is a broad-spectrum herbicide. In other words it is an herbicide that kills various plant varieties. Glyphosate is non-selective in terms of the plants it affects. Crop plants can be genetically engineered to make them glyphosate-resistant. In these cases glyphosate is used to protect the genetically modified plant, while all other plants are killed off. For years glyphosate has been the most widely used ingredient in

[1]https://en.wikipedia.org/wiki/Glyphosate.

J. Unterweger (✉)
Dr. Josef Unterweger, Rechtsanwalt, Buchfeldgasse 19a, 1080 Wien, Austria
e-mail: office@unterweger.co.at

L. Westra et al. (eds.), *The Role of Integrity in the Governance of the Commons*,
DOI 10.1007/978-3-319-54392-5_15

herbicides worldwide. In 2014 around 720,000 tonnes of glyphosate were used globally. In Germany, the figure was around 6000 tonnes.[2]

Monsanto patented glyphosate in 1970 and is one of the world's leading producers of the chemical. The company also has patents for genetically modified glyphosate-resistant plants such as Roundup Ready soybeans and Roundup Ready rape. Glyphosate accounts for around 40% of Monsanto's revenue in Germany.[3] In 2011 the company recorded a net profit of 1.6 billion dollars on revenues of 11.8 billion dollars. Around 27% of its revenue comes from herbicide production and sales.[4]

Critical journalists have called Monsanto one of the biggest polluters in industrial history, pointing to a remarkable string of scandals and a number of convictions. Monsanto's history is tied to the sweetener aspartame, polychlorinated biphenyl or PCB, contamination of the area surrounding the Monsanto plants in Anniston in Alabama, the production and distribution of Agent Orange, the chemical spill at Times Beach, Missouri, the bovine growth hormone Posilac, and the production and patenting of genetically modified plants. The company's history is also dotted with accusations of manipulation and corruption, some of which ultimately led to court convictions. Scientists report being pressured after publishing unfavourable papers.

A decisive point in this context is that some studies of glyphosate commissioned by Monsanto resulted in convictions for scientific fraud. For example, Industrial Biotest Laboratories (IBT Labs), which was contracted by Monsanto among others to conduct studies of glyphosate, was closed down by the US Justice Department in 1978.[5]

---

[2]Since 1970 when Monsanto the US chemical company filed their patent for the original Roundup herbicide, it has been a wholesale success all over the world. During the past year more than 720,000 tonnes were used worldwide, more than any other herbicide. 6000 tonnes ended up on fields in Germany—for weed control in grain and corn cultivation as well as in vineyards and orchards. To make matters worse, the Germany Railway system uses the herbicide to clear undesirable plants from their tracks (see article July 15, 2015). http://www.welt.de/wirtschaft/article144015187/Wie-gefaehrlich-ist-C3H8NO5P-wirklich.html. http://www.wallstreet-online.de/nachricht/6561010-global-glyphosate-market-is-expected-to-reach-usd-8-79-billion-by-2019-transparency-market-research.

[3]junge welt 1/06/2016/Topic/p. 12. Profitable Herbicide: Worldwide Glyphosate is used more and more. The authorities seem unconcerned that experts have announced that the herbicide is »probably carcinogenic "Peter Clausing". Glyphosate plays a threefold key role: First the chemical is extremely significant for herbicide producers—in Germany alone 40 per cent of Monsanto's sales are due to the sale of herbicides containing Glyphosate. https://www.jungewelt.de/2016/01-16/054.php?sstr=glyphosat.

[4]During the 2011 business year Monsanto turned over 11.8 billion US $, a net gain of 1.6 billion US-$. The *Agricultural Productivity* division produces herbicides for agriculture, industry, public maintenance, homes and gardens and accounted for 27% of turnover. The best known product of this division is the broad-spectrum herbicide Roundup. https://de.wikipedia.org/wiki/Monsanto.

[5]Marshall, E. (1983) "The murky world of toxicity testing", Science. 220 (4602): 1130 – 1132. doi:10.1126/science.6857237, PMID 6857237; Schneider, Keith, "IBT – Guilty" Winter 1983. Amicus Journal. Planetwaves.net; Schneider, Keith "Faking it: The Case against Industrial Bio-Test Laboratories", Amicus Journal, Spring 1983; https://en.wikipedia.org/wiki/Monsanto_legal_cases#Roundup.

In 1991 the owners of Craven Laboratories were indicted and convicted. Monsanto explained that the glyphosate studies concerned had been repeated, and that EPA certification of Roundup was no longer based on the studies conducted by Craven Laboratories and IBT Labs.[6]

The European Union's approval procedure is set up in such a way that approval of a pesticide in one member state is valid in all of the other EU countries. Pesticide producers are free to choose the member state to which they submit an application for approval. This means that the manufacturers can basically select any regulatory authority. They can also specify that the application documents remain confidential.

Applicants must provide evidence that their pesticide is not carcinogenic. According to Regulation EC number 1272/2008 a substance is classified as carcinogenic if at least two studies produce positive results.

EC Regulation number 1107/2009 establishes the precautionary principle and a high level of protection.[7]

Recital 8 of the Regulation reads as follows: The purpose of this Regulation is to ensure a high level of protection of both human and animal health and the environment and at the same time to safeguard the competitiveness of Community agriculture. Particular attention should be paid to the protection of vulnerable groups of the population, including pregnant women, infants and children. The precautionary principle should be applied and this Regulation should ensure that industry demonstrates that substances or products produced or placed on the market do not have any harmful effect on human or animal health or any unacceptable effects on the environment.

These objectives are reinforced by recital 24 of the same Regulation: The provisions governing authorisation must ensure a high standard of protection. In particular, when granting authorisations of plant protection products, the objective of protecting human and animal health and the environment should take priority over the objective of improving plant production. Therefore, it should be demonstrated, before plant protection products are placed on the market, that they present a clear benefit for plant production and do not have any harmful effect on human or

---

[6]*EPA FY1994 Enforcement and Compliance Assurance Accomplishments Report (PDF). United States Environmental Protection Agency;* https://en.wikipedia.org/wiki/Craven_Laboratories.

[7]Regulation (EC) No 1107/2009 of the European Parliament and of the Council of 21 October 2009 concerning the placing of plant protection products on the market and repealing Council Directives 79/117/EEC and 91/414/EEC; http://eur-lex.europa.eu/legal-content/EN/TXT/HTML/?uri=CELEX:02009R1107-20140630&qid=1475740883425&from=EN.

Plant protection products may only be authorised if: "industry demonstrates that substances or products produced or placed on the market do not have any harmful effect on human or animal health or any unacceptable effects on the environment". (Regulation (EC) No 1107/2009 concerning the placing of plant protection products on the market and repealing Council Directives 79/117/EEC und 91/414/EEC, recital 8). "The provisions governing authorisation must ensure a high standard of protection. In particular, when granting authorisations of plant protection products, the objective of protecting human and animal health and the environment should take priority over the objective of improving plant production". (Regulation (EU) No 1107/2009, recital 24).

animal health, including that of vulnerable groups, or any unacceptable effects on the environment.

If the regulatory authority confirms that the requirements for approval have been met, its decision is reviewed by the EFSA and then passed on, together with a draft decision, to the relevant committees of the member states, which then make a final decision. Thus the member states take the decision on approval, and if they cannot agree, the EU Commission takes responsibility.

In order to secure an extension of the authorisation, in May 2012 Monsanto submitted an application to the BfR, which had originally approved the substance in 2002 on behalf of a number of glyphosate producers and retailers. The current BfR president, Professor Andreas Hensel, was involved in the authorisation process. In its draft assessment report published at the end of 2013, the BfR had classified glyphosate as non-carcinogenic. Then, on 20 January 2014 the BfR stated that there were no indications that glyphosate was carcinogenic, toxic to reproduction or embryotoxic, adding that "there was no evidence of carcinogenicity up to the highest dose level".

In a press release published on 20 January 2014 the BfR explained: "In addition to the documents already incorporated in the first test series of active ingredients, more than a thousand new studies were examined and evaluated. These new studies do not suggest that glyphosate has carcinogenic or embryo-damaging properties or that it is toxic to reproduction in test animals. The data do not warrant any significant changes in the limit values of the active ingredient, says Professor Dr. Dr. Andreas Hensel".

On 20 March 2015 the International Agency for Research on Cancer or IARC, part of the WHO, the World Health Organisation, classified glyphosate as "probably carcinogenic to humans".

On 31 March 2015 the BfR published its final assessment report in which it contradicted its original finding that there was "no evidence of carcinogenicity" and noted a slight increase in the occurrence of tumours. However, it pointed out that the "slight increase in the incidence of malignant lymphoma" this was "not statistically significant". No reasoning was given to back up this conclusion.

The IARC commissioned some of the world's top scientists to evaluate glyphosate—the world's most widely used pesticide. The evaluation was performed by scientists who have already published research into glyphosate and other pesticides, and who are recognised in scientific circles as experts on this substance. The finding that glyphosate is probably carcinogenic to humans reflects a consensus among the scientists involved. The statement of 20 March 2015 classifying glyphosate as "probably carcinogenic to humans" was included in an IARC monograph published in July of last year.

Of the five studies on mice that the BfR deemed to have produced negative results in its interim report, two were examined by the IARC, and the Agency's monograph states that the studies provided sufficient evidence for carcinogenitity. In response the BfR released an addendum to its final assessment report on the pesticide on 31 August 2015. However, in the addendum the BfR concluded that there was a "dose-dependent, statistically significant increase" in malignant lymphoma. The

BfR explained that it had relied on industry data submitted by Monsanto in its findings. The Institute also said that the discrepancy between the opinions of the IARC and the BfR was down to data supplied by the glyphosate industry, adding that the statistical analyses of the industry and the IARC were both appropriate. At the same time, the BfR declared that the studies on which the IARC based its findings—all of which had been peer-reviewed and published—were "not reliable". Therefore the Institute concluded that glyphosate was not carcinogenic.

On 12 November 2015 the EFSA also classified glyphosate as non-carcinogenic, and as a result recommended its continued use in the European Union for the next 15 years.

A few days later, on 27 November Professor Christopher J. Portier sent an open letter to the EU's Health and Food Safety Commissioner Vytenis Andriukaitis, who is ultimately responsible for the approval of glyphosate. In his letter Professor Portier made serious accusations against the BfR and the subsequent evaluation by the EFSA. In a nutshell, the letter condemned the BfR's risk assessment as "scientifically unacceptable", "fundamentally flawed" and "misleading". In his open letter, Professor Christopher J. Portier and the co-signatories made serious accusations against the BfR and the subsequent evaluation by the EFSA. The authors explained that "the BfR decision is not credible because it is not supported by the evidence", adding that the BfR's conclusions were misleading, as was its language, which in turn was not internationally acceptable and as a result failed to meet EU Guidelines.

Professor Portier continued that "it is clear that BfR differed from standard scientific practices in order to reach their conclusions". He also accused the Institute of inappropriate use of historical data. The scientists concluded that the studies "in fact document the carcinogenicity of glyphosate".

A further accusation was that the BfR used testing guidelines to exclude substantive scientific evidence from its cancer risk assessment and also ignored OECD guidelines. The arguments put forward by the BfR were found to be "fundamentally and scientifically flawed". Furthermore, according to Professor Portier, the evaluations carried out by the BfR and the EFSA did not reflect the available science.

The BfR—the regulatory authority selected by the pesticide industry and pesticide producers—gradually changed its assessment of the studies submitted by the applicants. The original finding was that "there was no evidence of carcinogenicity up to the highest dose level" was replaced by a "slight, but statistically insignificant increase in the incidence of malignant lymphoma" and ultimately "a statistically significant increase in malignant lymphoma that could be seen as treatment-related". In spite of this opinion, the BfR stood by its finding that there were no indications of carcinogenicity.

In justifying these discrepancies and corrections, the BfR pointed out that it had originally based its assessment on statistical analyses and data provided by the glyphosate industry. The EFSA accepted BfR's incorrect assessment without hesitation.

GLOBAL 2000 asked toxicologist Peter Clausing and epidemiologist Eberhard Greiser to prepare an expert opinion on the evaluation of the studies submitted by the glyphosate industry. Mr. Clausing and Mr. Greiser confirmed the accusations made in the open letter. They found that in the approval application, the industry claimed that there were no indications of carcinogenicity. However, the studies that accompanied the application, which were intended to prove that glyphosate was harmless, actually showed a rising incidence of tumours at increasing dosages of the substance. It further emerged that the application for extension of the approval was based on animal studies that had not been properly or professionally evaluated and interpreted. As a result, significant carcinogenic effects that appeared in all five of the studies submitted by the industry were concealed.

Regulation 1272/2008[8] states that carcinogenic pesticides must not be approved for use in the European Union. A substance is classified as carcinogenic if at least two studies produce positive results for cancer. The approval application submitted by the glyphosate industry was based on five studies of carcinogenicity in laboratory mice. In all five, tumours developed in the kidneys, blood veins or lymph glands.

Monsanto has been linked with scientific fraud and falsification of studies several times. It even has a number of convictions in connection with studies of glyphosate. Although it was aware of these circumstances, the BfR accepted the data and conclusions provided by the glyphosate industry in the approval application without the necessary review. The Institute only reviewed the cancer studies submitted by Monsanto after the IARC had classified glyphosate as probably carcinogenic to humans.

In spite of these results, the BfR did not withhold approval, but instead continued to adopt the position put forward by the pesticide industry. After the open letter had indicated yet again that incorrect conclusions had been reached, and statistically significant carcinogenic effects had been obscured, the BfR not only revised the obviously incorrect findings, it also ruled in favour of extending the authorisation for glyphosate when this clearly ran counter to the facts.

In spite of the evidence of shortcomings in the BfR's assessment, EFSA accepted their arguments and conclusions and contradicted the available study results by announcing that glyphosate was not carcinogenic. EFSA recommended its continued use in the European Union for 15 years. It is unusual for complaints to be filed against public bodies such as the BfR, and its unusual nature will be shown a little later.

On 2 March 2016 GLOBAL 2000 and seven other environmental organisations filed complaints against Monsanto on, which had overall responsibility for the glyphosate

---

[8]Regulation (EC) No 1272/2008 of the European Parliament and of the Council of 16 December 2008 on classification, labelling and packaging of substances and mixtures, amending and repealing Directives 67/548/EEC and 1999/45/EC, and amending Regulation (EC) No 1907/2006 (Text with EEA relevance), 3.6. Carcinogenity; http://eur-lex.europa.eu/legal-content/EN/TXT/?qid=1475741055994&uri=CELEX:02008R1272-20160401.

approval application, and against the BfR and the EFSA both in Austria[9] and in Germany[10]. The complaints were based on suspicion of serious commercial fraud, because incorrect or falsified data had been used to generate profit, thus giving consumers and businesses the erroneous idea that the pesticide was not harmful to human and animal health or to the environment, as provided for in EC Regulation No. 110/2009.

The statement by the International Agency for Research on Cancer on 20 March 2015, the responses of the BfR, and Professor Christopher J. Portier's open letter to the EU Commissioner for Health and Food Safety, also signed by 96 renowned scientists, have brought to light a number of extremely suspicious circumstances. Expert opinions commissioned by the environmental organisations have confirmed and also strengthened these suspicions, as have the BfR's reactions, especially those of the Institute's president. The BfR president had made unilateral comments favouring the pesticide industry by stating: "Glyphosate has been used in agriculture for over 40 years and there has never been any serious evidence of damaging side-effects". In addition, his statement contained misleading advertising slogans used by Monsanto.

GLOBAL 2000 filed another complaint on 20 April 2016 after an expert opinion disproved the BfR's claims once again. The Vienna public prosecutor's office demanded that Monsanto respond to the complaint. Monsanto interceded and demanded to bar GLOBAL 2000 from access to the files. In March 2016 the prosecutor gave access to the files but revised this decision in July 2016. In October 2016 the prosecutor considered to investigate fraud and intended to nominatae an expert witness. In November 2016 the prosecutor dismissed the complaint and refused to give grounds for the dismissal.

For 3 months the Berlin public prosecutor's office was unable to confirm receipt of the complaints, and in late June 2016 it stated that they had been passed on to the Düsseldorf public prosecutor's office for reasons of jurisdiction. The prosecutor decided more than 5 months after submission of the complaint to dismiss the complaint without investigation.

So what was the outcome of the environmental organisations' efforts? The vote by the EU member states on extending glyphosate's approval was postponed several times, and even a third vote on 6 June this year did not produce a qualified majority in favour of an extension. However, the European Commission has extended the approval for glyphosate until the end of 2017.

Several studies have shown that the serious accusations against the BfR and the EFSA made by Professor Christopher J. Portier in his open letter are correct. As a result of the complaints filed in Vienna, Monsanto will be obliged to respond to the accusations.

Another outcome was that in April 2016 the environmental organisations had problems finding a room for a press conference held after the initial complaint had been submitted. On three successive occasions, after revealing the subject of the

[9]Case Nr: Staatsanwaltschaft Wien 32 St 17/16z.

[10]Case Nr: Staatsanwaltschaft Düsseldorf 10 UJs 1993/16.

press conference, the organisations were refused access to rooms for which they had confirmed bookings.

A further outcome is that the European Union's approval procedures for pesticides could be overhauled. The goal should be to establish an independent EU-wide authority, a step that would end the "race to the bottom" among the national regulatory authorities. The approval process should be transparent and science-based, and should be subject to independent supervision. The public should also play a part in the process. Explaining the precautionary principle is not enough—it also needs to be applied.

At present, the BfR is taking decisions that may impact the health of 500,000,000 EU citizens and their children. Belgians, Hungarians, Italians, French and British will be affected by the BfR's decisions, but have no say in the matter. And that needs to change.

1 Addendum: "Salt"

The BfR responded swiftly to the complaints brought by the environmental organisations. A study by environmental associations found glyphosate residues in beer and foodstuffs, as well as in urine samples taken from a large number of people.

BfR president Andreas Hensel responded by saying: "Glyphosate has been used in agriculture for over 40 years and there has never been any serious evidence of damaging side-effects".[11] This comment was made on 12 March 2016—after the IARC had classified glyphosate as probably carcinogenic to humans and had published the corresponding monograph. Professor Christopher J. Portier's open letter to EU Commissioner Andriukaitis contained some serious accusations against the BfR.

Professor Hensel explained that "the lethal dose of glyphosate is comparable to that of table salt".[12] This statement is astonishing given that Monsanto used this comparison to advertise glyphosate in the US.

In 1996, the Attorney General of the State of New York Consumer Frauds and Protection Bureau and Environmental Protection Bureau instigated proceedings against Monsanto based on this and similar advertising slogans. That same year, Monsanto was obliged to cease and desist from making statements such as "Glyphosate is less toxic to rats than table salt following acute oral ingestion".[13] "Pesticide

[11]http://www.spiegel.de/spiegel/vorab/behoerdenchef-wirft-umweltverbaenden-und-gruenen-panikmache-vor-a-1081815.html (German only).

[12]http://www.spiegel.de/spiegel/vorab/behoerdenchef-wirft-umweltverbaenden-und-gruenen-panikmache-vor-a-1081815.html (German only).

[13]http://www.mindfully.org/Pesticide/Monsanto-v-AGNYnov96.htm; Attorney General of the State of New York. Consumer Frauds and Protection Bureau. Environmental Protection Bureau. 1996. In the matter of Monsanto Company, respondent. Assurance of discontinuance pursuant to executive law 63(15). New York, NY, Nov. False Advertising by Monsanto Regarding the Safety of Roundup Herbicide (Glyphosate).

products containing glyphosate or any component thereof are safe, non-toxic, harmless or free from risk" and "glyphosate-containing pesticide products or any component thereof are safer or less toxic than common consumer products other than herbicides".

2 Addendum: "Unreliable Studies"—Professor Eberhart Greiser

In light of the criticism of the BfR sparked by Professor Portier's open letter in November 2015, the environmental organisations asked German epidemiologist Professor Eberhart Greiser of the University of Bremen to provide an expert opinion on the way BfR dealt with human evidence. The BfR had dismissed numerous peer-reviewed studies as unreliable that had appeared in respected journals. The open letter described this as "fundamentally flawed", "scientifically unacceptable" and "misleading". Some of the studies were mentioned by the pesticide producers in the approval application, which alleged that they contained methodological errors, even though they were peer-reviewed. The BfR accused eighteen epidemiological studies of failing to collect the necessary information on the risk of disease. Professor Greiser looked into this accusation and concluded that all of the studies had been published in renowned medical journals and peer-reviewed. Three of the studies described as "inconclusive" by the BfR were carried out by the National Cancer Institute in the US. Professor Greiser found that all of the studies contained the requisite information on the risk of disease—the exact opposite of the BfR's claims. While the BfR asserted that key data—such as exposure to glyphosate, smoking behaviour and previous illnesses—had not been collected, the review showed that all of this information had actually been compiled in detail, in line with the latest epidemiological methods.

In view of the statements made by the BfR and in particular by its president, as well as the scientists' conclusions in the open letter, the environmental organisations filed a supplementary complaint with the Berlin public prosecutor's office, because it had been shown that the BfR president had unilaterally made comments favouring the pesticide industry.

The BfR's comments that studies on the health-related effects of glyphosate were "unreliable" and "irrelevant" were obviously off the mark. These comments were also only made in connection with studies that identified the health risks posed by glyphosate. The supplementary complaint was submitted in view of all of these facts.

Laudato Sì and the Christian Ecological Utopia

Philippe Crabbé

1 Introduction

Following the German sociologist Karl Mannheim, Quebec sociologist Fernand Dumont defined a utopia as being an ideology which motivates people to change an existing situation and, therefore, is oriented towards the future. An ideology is a synthetic and systemic narrative aimed at motivating its readers or listeners to action. This narrative is not a scientific theory but is made of scientific "residuals". It is thus compatible with science (Dumont 1974). Sustainable development is such a utopia (Crabbé 1997). Eco-theology is another as well, as the latter certainly motivates the religious believer towards action, is not scientific but is altogether compatible with scientific theories.[1] Some writings of Pierre Teilhard de Chardin, S. J., are of this nature. After forbidding Teilhard from teaching, the Catholic Church since John Paul II has gone a long way towards the former's rehabilitation.[2] The Catholic Church was slow in changing its Aristotelian/Ptolemaic geocentric view of cosmology, of immanent finality, until the middle of the nineteenth century (Nys 1908); Lutherans have been exploring new cosmological venues since the beginning of the 19th century (Lyons 1982).

---

[1]The distinction between eschatology and utopia is pertinent here. However, a discussion of this distinction would lead us too far afield. Suffice it to say that in Dumont's definition of utopia, the word "imaginary" never appears. I am indebted to Ron Engel for pointing out this distinction to me and for other valuable comments. However, remaining errors are mine alone.

[2]His rehabilitation by the Vatican began in 1981 with a letter, on behalf of Pope John Paul II, from Cardinal Casaroli, Secretary of State of the Vatican, to the Rector of the Institut Catholique de Paris. http://www.sj2014.net/blog/12-may-1981-important-letter-rehabilitates-teilhard-de-chardin retrieved March 4 2017.

P. Crabbé (✉)
Institute of the Environment, University of Ottawa, Ottawa, ON, Canada
e-mail: crabbe@uottawa.ca

L. Westra et al. (eds.), *The Role of Integrity in the Governance of the Commons*,
DOI 10.1007/978-3-319-54392-5_16

The concept of "Cosmic Christ" appeared in Christian theology at the beginning of the nineteenth century, especially in Germany and therefrom in Great-Britain. The concept emerged in the United-States at the beginning of the twentieth century and, in France, roughly since the First World War, in Teilhard de Chardin's work (Lyons 1982, Appendix). It is noteworthy that Teilhard de Chardin studied theology in Great-Britain from 1908 to 1912 (Lyons 1982). Joseph Sittler, a Lutheran theologian, is credited with the recent revival of the concept in the United-States (1954) and he claimed to rely on Galloway (Galloway 1954), a Scottish Evangelical theologian.

Eco-theology is always conditioned by a worldview, by a cosmology. "...Right from the beginning of the history of Christology, the understanding of Christ has been determined not merely by sources of divine revelation but also by man's conception of the cosmos" (Lyons 1982, p. 12). In the Middle Ages, a certain form of dualism was introduced by Thomas Aquinas. Only humans benefit from grace and, therefore, humans are superior to the rest of nature. With the advent of modernity and Kant's influence, cosmological physical phenomena were expected to belong to the realm of sciences while mankind's destiny and the study of moral values belonged to philosophy and faith (Galloway 1954). Progressively, a cosmocentric worldview attempted to take over from anthropocentrism in Christian eco-theology but did not quite succeed by fear of pantheism.

This paper begins by examining the Catholic Church ecological teaching before Pope Francis encyclical *Laudato Sì* in order to determine the latter's originality. A second section identifies what's original in the encyclical despite its endorsement of previous teaching. A third section provides a summary of Christian eco-theology as, for this author, there is little variation on the subject among the various Christian denominations. Finally, a fourth section introduces the concept of "Cosmic Christ", not mentioned in Francis' encyclical. A concluding section underlines the almost complete ignorance of Christian eco-theology among the Catholic faithful.

2 The Catholic Church Ecological Teaching Before "Laudato Sì"

Laudato Sì builds upon the ecological teachings of the Council Vatican II in the nineteen sixties and on the ones from Francis' predecessors, from Paul VI onward. Mankind's stewardship and not ownership over creation is well-established in these teachings. This stewardship consists in submitting nature, in perfecting it for the development of the whole of mankind, under God's authority. The Catholic Church calls for international governance "to work for the solution to the serious problems of our times" (*Gaudium et Spes*, 1965, [34.1, 57.2, 84.3][3]).

[3]Figures between brackets refer to the paragraph and sub-paragraphs in the original document.

Pope Paul VI is the first of the post-second world war popes to have addressed ecology directly. He warned in a speech to the Fao on the occasion of its 25th anniversary in 1970 that time has come for mankind to "dominate its domination" [4] and to embrace "integral development" (Populorum Progressio, 1967) anchored in human solidarity, i.e. technological and economic progress accompanied by moral and social progress [4]. Opposition to artificial birth control as a mean of population control is reaffirmed [6]. In a 1972 message to Maurice Strong on the occasion of the UN Conference on the Human Environment, Paul VI emphasized that the ecological interdependence between the environment and mankind and among the various countries requires joint responsibility and common destiny by solidarity; technological solutions do not suffice. The environment is the common patrimony of mankind. Therefore, mankind must control its material growth, use the earth's food with wise moderation, and cultivate real poverty of spirit as did Saint Francis of Assisi. Paul VI warns against depriving less industrialized countries of their development means while these countries already carry a heavy burden, having to face degradation of the common environment (the ecological debt caused by industrialized countries).

Pope John-Paul II in his important message for World Peace Day 1990, entirely devoted to ecology, lays out the sacramental approach to eco-theology [3] (see Sect. 3), says that environmental degradation, including climate change, is only one aspect of a moral crisis, i.e. the lack of respect for life as result of an unnatural and reductionist vision [6–7]. The human race is called to explore the ecological order, to examine it with due care and to make use of it while safeguarding its integrity [8]. The right to a safe environment must be a human right [9.2]. John-Paul makes the connection between environmental degradation and poverty which calls consumerism into question [11, 13]. "An education in ecological responsibility is urgent" [13]. "Finally, the aesthetic value of creation cannot be overlooked.... Good urban planning is an important part of environmental protection," [14]. Responsibility for and duty towards nature is an essential part of the Christian faith; as a result, there is a vast field for ecumenical and interreligious cooperation [15]. The example of Saint Francis will help us keep alive a sense of "fraternity" with all those good and beautiful things which God has created. We have a serious obligation to care for them, "in light of that greater and higher fraternity that exists within the human family" [16.3].

John Paul II emphasized the priority of ethics over technology and reiterates that one has to protect the moral conditions of an authentic "human ecology", a rewording of Paul VI's integral development (Redemptor hominis, [16.1], 1979; Centesimus annus, [38.1], 1991). These comments apply less to an economic system than to an ethical and cultural system which may be reduced to the production of goods and services. The State must protect the common goods that are the natural and the human environment that cannot be protected solely by the market. While the old capitalism had to protect labor, the new capitalism must protect the common goods. While praising the market, John-Paul II wanted to limit its purview (Centesimus annus, [40, 42], 1991). "We need to realize therefore that there can be a grave sin against the natural environment", citing a document from the Polish bishops (Homily at Zamosc, Poland, [3.4], 1999). He offered a partial summary of the prophetic (eschatological) approach to creation theology (see Sect. 3; ibid. [4.1]). He introduced the concept of "ecological conversion" (*L'engagement pour éviter une crise écologique majeure*, Audience générale, 17 janvier, [4], 2001).

Benedict XVI noted that the "creation doctrine" had almost disappeared from theology and, citing Saint-Paul's Epistle to Romans chapter 8, said that creation and Redemption, terrestrial life and aeternal life, responsibility towards creation and responsibility towards other human beings and the future must go hand in hand (Meeting with the clergy of the diocese Bolzano-Bressone and answer to a question from Karl Golser, 2008). He said: "The Church has a responsibility towards creation" and she must assert this responsibility in the public sphere. He underlined "the profound significance of the culture of different nations" without falling into relativism (eclecticism) and the danger of cultural levelling. What eclecticism and cultural levelling have in common is the separation of culture from human nature." (Caritas in Veritate, [26], 2009).

3 Pope Francis' Encyclical "Laudato Sì"

From the short review of the Catholic Church teaching on ecology before Pope Francis, it is clear that the major tenets of Pope Francis' ecological conversion doctrine were already in place.

Pope Francis is the first pope to have devoted an entire encyclical to the theme of ecology, however. The encyclical's objective is to "dare transform into personal suffering what's occurring in the world" [19]. In other words, the objective is to extract ourselves from our indifference towards the poor and the state of creation which are related to each other.

Francis knows that public opinion is divided on the existence of the ecological crisis or, at least, on its remedies. Sending back to back cornucopians (relying on technological change) and malthusians (relying on population control), he believes that the ecological crisis has a solution and that it lies in ecological conversion through education. This ecological conversion requires first a cultural and ethical conversion. One must abandon the domination culture over the other and over nature and replace it by one of fraternity with both. A behavioral change, not only individual but collective, is required through which one must abandon consumerism to the benefit of sobriety, the exploitation of nature to the benefit of its protection, the utilitarian perspective to the benefit of universal solidarity. Technological change does not suffice in performing the ecological conversion. The encyclical endorses the ethics of Saint-Francis of Assisi, i.e. the one of fraternity with people and with the whole of nature, which has intrinsic value. Education is the tool required to spread this alternative ethics.

Fundamentally, the encyclical offers little which is new. What is new is its tone of radicalism and urgency. The encyclical reinforces previous papal teaching by being entirely devoted to ecology and in rejecting all forms of domination, considered as sins. Ecology from being peripheral has now become central to the Catholic Church message.

What's new in Laudato Sì is first the use of the Trinity of God, a set of relations,[4] as the model for the ecological conversion and as part and parcel of what Francis calls "integral ecology", i.e. the ecology which deals with people's relations—including their moral dimensions—among themselves and with the natural world, i.e. "human ecology", and with the Trinity. Integral ecology may be considered as radical ecology to the extent that it bans all forms of domination considered as sin, utilitarian ethics and, to some extent, anthropocentrism about which it is more hesitant (Crabbé 2016). Integral ecology is an echo of Paul VI's "integral development".

A second originality of the encyclical is its major emphasis on interdisciplinarity, the needed dialogue among disciplines to which Francis adds religion.

A third one is its emphasis on the spirituality of Saint-Francis of Assisi who personified the natural world, granting it its intrinsic value, without falling into the pantheism trap.

A fourth one is that, despite its emphasis on ecology, Francis maintains that the objective of his encyclical is social. It is social because: (1) mankind is part of nature; (2) mankind carries a responsibility towards nature, to develop it, to protect it and to lead it back to its creator; (3) behaviors towards the natural world affect behaviors towards humans and vice versa.

A fifth one is that Francis offers a very comprehensive post-modern critique of the concept of human progress without sinking into relativism. Technology is a determinant of social relations and culture. The techno-economic paradigm orients consumers' preferences but deprives consumers from their freedom. The pope rejects mass production and consumerism, and merchandising as applied to common goods such as carbon, water, etc. He also rejects reductionism (especially in relation to technological progress) and all forms of domination and power structures among which he includes economists, financiers and multinationals.

A sixth one follows from the previous one, the indictment of capitalism in contrast with its relative endorsement by John Paul II.

A seventh one is that Francis wishes his encyclical to be collegial and interreligious. It is collegial to the extent he quotes his bishops all over the planet. It is interreligious to the extent he quotes extensively Orthodox Patriarch Bartholomew and refers to the Earth Charter and to a Sufi master but ignores completely the protestant contributions, including the one which clearly influenced the theology of his encyclical, Jürgen Moltmann's (Crabbé 2016).

A eighth one is the condemnation of domination of people and nature as a sin, condemnation allegedly borrowed from Orthodox Patriarch Bartholomew but the needed "domination of domination" was already present in Paul VI's writings.

---

[4]This theme is developed in Jürgen Moltmann's theology of creation. Moltmann uses various concepts from systems theory that the pope adopts in his encyclical.

An ninth one is his relatively systematic coverage of creation theology, based on the interpretation of the biblical texts (the hermeneutical approach) pointing towards mankind's stewardship, on creation as divine gift (sacramental approach), and on the ultimate divinisation of creation through the continuity of the creative action, for which humans are expected to be "co-creators" (the prophetic or eschatological approach). The pope endorses resolutely the theology of creation as a whole, as well as Teilhard de Chardin's perspective on creation and on the human responsibility in creation. He never refers to "Cosmic Christ", however, despite mentioning Teilhard de Chardin in a footnote full of caveats. Humans have the cosmic responsibility to take creation home to its creator.

An tenth one is that Francis acknowledges the signs of the times[5]: Sustainable development and UN conferences on this topic and on climate.

The encyclical offers a good assessment of what current science tells us about the current state of the natural world under climate change and of the interdependencies between the natural world and society.

The encyclical may not be an innovative text but it is a forceful one. Published in 2015, one could have expected a leap forward with respect to similar documents published 45 years ago. This is not the case except for the response to a challenge which hardly existed 45 years ago and that Francis courageously met in his encyclical, the challenge of climate change. He resolutely endorses the precautionary principle and the one of Common but Differentiated Responsibility, which are ethical principles.

The American Baptist Churches produced a document titled "Creation and the Covenant of Caring" in 1989, which is a model of clarity and concision (4 pages!) and which proclaims substantially the same message as Laudato Sì (196 pages!). According to H. Paul Santmire, the American Lutheran Church published a statement "The Human Crisis in Ecology" in 1972. The Unitarian Universalist Association's "Public Policy Statement on the Environment" dates from 1971 (Crabbé 2016).

4 Christian Eco-Theology[6]

Though I am no theologian, it is my reading, confirmed by the Lutheran theologian Jürgen Moltmann, that there is little difference among the catholic, the orthodox, the reformed (and the Jewish) eco-theologies. If I am right, this is a further reason

[5]The signs of the times are the phenomena which, by their expanse and frequency, characterize an epoch and the concerns of mankind about them (Crabbé 2016, fn. 28).

[6]Many thanks to Fr. Emmanuel Durand, o. p., Dominican University College, Ottawa for his suggested improvements to this section. Remaining errors are mine. This section is mainly based on Moltmann (2004) and Egger (2012).

for the three churches to cooperate in disseminating the Christian ecological utopia. The latter is a nice but underutilized story to motivate changes in behavior.

Judaïsm, having rejected in the Old Testament the idea of another world being promised by God as a result of the Alliance with Noah, had no choice but to extend to the entire universe the promise of Redemption offered by God. This is the ultimate (eschatological) meaning of the Promised Land in the Jewish apocalyptic literature (Galloway 1951).

In his encyclical letter Laudato Sì (chapters 2 and 6), Francis develops the catholic eco-theology. It is very similar to the one developed by the Lutheran Jürgen Moltmann and is enriched as well as Moltmann's by Orthodox theology. The latter is essentially grounded on the Church Fathers, i.e. on the common patrimony of all the Christian Churches. According to Moltmann, it may be summarized by one word: panentheism, i.e. God is present in nature and nature is destined to rest in God (Moltmann 2004).

According to Francis, human existence rests upon three fundamental relations: the relation with God, the one with humans, and the one with the rest of nature. These relations were broken by sin with the result that these three relations have become conflictual [66].

There are three approaches to eco-theology. The first one is based upon the hermeneutics of biblical texts. It is clear that the Bible as a whole identifies God as the master of creation and that God did not abandon this power to mankind. Humans are only stewards whose role is to make creation bear fruit, to protect it and not to do with it whatever they pleased. The rule of the Shabbat applies to humans as well as to the rest of nature.

The second approach, dear to Saint-Francis of Assisi and to the Orthodox, is based on creation being the gift of nature from God to humans. Creation is "good" in the eyes of God and this divine gift must be respected by humans; it has intrinsic value. As Pope Francis says, creation's "being" is more than its usefulness to humans [69]. As a gift of God, creation is a sacrament, i.e. the visible expression of God's love which assigns its value to the visible object. Moreover, God inhabits his creation. Creation is a common heritage meant to benefit all of mankind.

Respect for creation consists not only in contemplation but also, in the spirit of the parable of the talents, in transforming the former through work before returning it to its creator. In exchange, mankind must protect it. Creation needs continuing divine creative action and the development of its potentialities. If God doesn't need us to be present in the natural world, creation needs human beings in order to realize its spiritual potential, manifest in fullness its sacramentality, lead to term its painful delivery, achieve the transfiguration project which was instilled in it by God from the origin.

The two first approaches can be found in Paul VI teachings. There is a third approach to eco-theology, the one of the cosmic dimension of salvation that one can find in John Paul II Homily at Zamosc. The salvation promise was made to the entire creation. Since the beginning of the world, but especially since his Incarnation, Christ's mystery operates, secretly, in the whole natural world, without affecting the latter's autonomy. Creation and Redemption go hand in hand. While bearing labor pain in union with Christ crucified, mankind collaborates with Christ to its own Redemption. The coming here and now of the "new heaven" and the "new earth" promised by the book of Revelations (Rev 21, 1) depends on us too. According to this third ecological approach to eco-theology, the bodies of the saved are expected to be transfigured, i.e. to participate in God's life—in the same way Jesus was transfigured in front of some of his disciples—in order to become the place of "rest" (Sabbath) of God. Francis says: "Sunday is the day of the Resurrection, the "first day" of creation, whose beginnings are mankind resurrected by Christ, token of the final transfiguration of the whole created reality. "[237] Moreover, this day announces "the eternal rest of man into God" [237]. Yes we are travelling towards the Sabbath of eternity, towards the new Jerusalem, towards the common home of heaven" [243].

Creation—each creature keeping its individuality—is destined to become the transfiguring inhabitation of God Trinity who will transform the former into a new heaven and a new earth, the kingdom of God. The whole cosmos participates in Christ of the mystery of Trinity: God the Father creates the world in view of the incarnation of his Son and in order to provide a cosmic body to this Son. The Spirit leads all creatures to their accomplishment. This is the interpretation of the Franciscan theologian Duns Scot (thirteenth century) and of Eastern Christianism. The Eucharist is a realization of this transfiguration because it deifies bread and wine. At the end of times, creation is called upon to be deified, i.e. to be in God. Francis says: "The divine Persons are subsistent relations, and the world created according to the divine model is a net of relations". [240].

This third approach to creation mays itself be the subject of three interpretations.

The first one, pessimistic, is the one of the *Dies irae*. Creation is essentially a temporary prison and there is no reason to linger in it.

The second is the one common to Catholic and Reformed churches. Through its sin, mankind lost the earthly paradise and precipitated in its fall the whole of creation even if non-human nature has not sinned. Through his incarnation, Christ redeemed the entire cosmos and the latter will recover its primeval form of earthly paradise.

The third one, the one of Moltmann and of the Eastern Church fathers—and thus orthodox—, is the divinization of the cosmos. Christ has not come only to redeem the world from sin but also to perfect the alliance with creation. Even without the Fall, Incarnation would have occurred. The Church fathers, in the Genesis stories, look less for a description more or less explanatory of the "beginnings" of the Universe than for drawing a vision of the world to come, of what God wishes creation to tend to and that will be realized at the end of times.

For Eastern Christianity, man is a microcosm, at the same time content and recipient of the cosmos, the bridge between heaven and earth. Non-human creation is also endowed with a mediation role between God and the human being: it is through the Eucharistic elements of creation, of which bread and wine are the symbols that human beings are in communion with God.

The human person is king, steward and liturgist of creation. She must be queen in the manner of Christ, i.e. be the servant of creation. As steward, he must empathize with the suffering of all creatures and understand their meaning, their divine objective. "Submit the earth" in this perspective is to transform the earth into a temple, said the orthodox theologian Serge Clément. As liturgist, the human person celebrates creation and loves it as oneself. All natural elements (water, plants, oil, wheat, incense, fire, wood, etc.) are integrated into the celebrations, offered to God and transformed through the Holy Spirit in places where he is present. For the Orthodox Church, sacraments don't have the sanctification of the soul and of the body of the faithful as purpose only, but also the divinization of creation. Francis says: "Indeed, the Eucharist is in itself an act of cosmic love. . .in the Eucharistic bread, 'creation tends towards its divinization'". [236].

"The relation of the human being as person with a personal God must also allow him to personalize the world" (Vladimir Lossky). This personal dimension allows to transcend the limits of matter and to come, as superior conscience, into relation with what is beyond the latter. This is why creation awaits with impatience the revelation of the sons of God (Paul, Rom 8, v 19) including its accomplishment inside us.

"The state of the cosmos demands not only that God becomes man but that man becomes god." (Olivier Clément) "Contemplation of nature as a way to the knowledge of God is a cardinal dimension of the orthodox mystique." (Egger 2012).

5 Christian Utopia: The Cosmic Christ

The concept of cosmic Christ has its roots in the New Testament, especially in Saint-Paul's epistles to the Colossians and to the Ephesians and in the Church fathers (Maloney 1968). The recent resurgence of the Cosmic Christ is due to the development of natural sciences in the nineteenth century, especially the geological and biological sciences (evolution). Thus the understanding of the role of Christ is tributary not only on Revelation but also in our understanding of our universe or cosmos (Lyons 1982). The universe is conceived more and more as a complex system, which makes it compatible with the immanence of God in his creation. Against idealism, Incarnation underlines the value of matter while, against materialism, it underlines the role of matter as agent through whom God reaches his spiritual objective. The reason for Christ's incarnation is not exclusively to redeem our sins but also, because of his cosmic nature, to perfect and unify the universe. Incarnation has, therefrom, a divine, human and cosmic dimension. The cosmic nature of Christ is directed towards the construction of his cosmic body. The Cosmic Christ doctrine eliminates Aquinas' dualism between nature and grace and between the Church and the world. For Teilhard de Chardin, Christ is the goal of biological evolution (ômega). The latter does not occur at random as claimed by biologists. The Church has not endorsed all the views of Teilhard de Chardin, especially not his blind acceptance of biological evolution with its concurrent rejection of the weakest, because God does not reject the weak. Through the

Cosmic Christ, eco-theology has become Christocentric and, therefore, cosmocentric (Moltmann 2004).

6 Conclusion

Pope Francis' Laudato Sì is firmly anchored in its predecessors' ecological teaching. The encyclical originality resides more in its tone than in its substance. Since the encyclical lagged behind the protestant ecological contribution by about 45 years, one could have expected more originality. However, the Christian eco-theology as it stands offers a utopia worth relying on to motivate the faithful to action. Unfortunately, the former is little used to that end in the Catholic Church at least.[7] Though Teilhard de Chardin has been rehabilitated to some extent by the Catholic Church, he remains a controversial figure among theologians and scientists alike. While the "Cosmic Christ" concept has been annexed by the New Age literature (Fox 1988), the Catholic Church has shied away from the concept because of this annexation (Fitzgerald 2003).

References

Benedict XVI (2008) Meeting of the Holy Father Benedict XVI with the clergy of the diocese of Bolzano-Bressanone, Cathedral of Bressanone, Wednesday, 6 August, Fr Karl Golser. https://w2.vatican.va/content/benedict-xvi/fr/speeches/2008/august/documents/hf_ben-xvi_spe_20080806_clero-bressanone.html. Retrieved 30 November 2016

Benedict XVI (2009) Encyclical letter Caritas in Veritate of the Supreme Pontiff Benedict XVI to the bishops, priests and deacons, men and women religious, the lay faithful and all people of good will on integral human development in charity and truth. http://w2.vatican.va/content/benedict-xvi/en/encyclicals/documents/hf_ben-xvi_enc_20090629_caritas-in-veritate.html. Retrieved 30 November 2016

Council Vatican II (1965) Pastoral Constitution on the Church in the modern world Gaudium et Spes promulgated by His Holiness, Pope Paul VI on December 7. http://www.vatican.va/archive/hist_councils/ii_vatican_council/documents/vat-ii_const_19651207_gaudium-et-spes_en.html. Retrieved 30 November 2016

Crabbé P (1997) Sustainable development: concepts, measures, market and policy failures at the open economy, industry and firm levels. Occasional Paper n° 16, Industry Canada, Ottawa

Crabbé P (2016) "Laudato si'", une responsabilité cosmique de la maison commune, Vertigo la revue électronique en sciences de l'environnement [On line], 16(2), September 2016. http://vertigo.revues.org/17692. Retrieved 30 November 2016

Dumont F (1974) Les idéologies. Presses Universitaires de France, Paris

[7]While "Laudato Sì" was published in several countries with comments (e.g. Belgium, France, U.K.), this does not seem to be the case in Canada and the U.S. The last two respective websites are sending their readers to documents which predated the encyclical. When is the last time you heard a homily on creation?

Egger M (2012) La terre comme soi-même. Repères pour une ecospiritualité. Labor et Fides, Genève
Fitzgerald ML (2003) Intervention, Pontifical Council for Culture, Pontifical Council for Interreligious Dialogue, Presentations of Holy See's document on "New Age". http://www.vatican.va/roman_curia/pontifical_councils/interelg/documents/rc_pc_interelg_doc_20030203_press-conf-new-age_en.html. Retrieved 30 November 2016
Fox M (1988) The coming of the cosmic Christ. Harper, San Francisco
Francis (2015) Encyclical Letter Laudato si' of the Holy Father Francis on care for our common home. http://w2.vatican.va/content/francesco/en/encyclicals/documents/papa-francesco_20150524_enciclica-laudato-si.html. Retrieved 30 November 2016
Galloway AD (1951) The cosmic Christ. Nisbett and Co, London
John Paul II (1979) Supreme Pontiff, encyclical letter Redemptor hominis to his venerable brothers in the episcopate, the priests, the religious families, the sons and daughters of the church and to all men and women of good will at the beginning of his papal ministry. http://w2.vatican.va/content/john-paul-ii/en/encyclicals/documents/hf_jp-ii_enc_04031979_redemptor-hominis.html. Retrieved 30 November 2016
John Paul II (1990) Message of his holiness Pope John Paul II for the celebration of the world day of peace. Peace with god the creator, peace with all of creation, 1 January. https://w2.vatican.va/content/john-paul-ii/en/messages/peace/documents/hf_jp-ii_mes_19891208_xxiii-world-day-for-peace.html. Retrieved 30 November 2016
John Paul II (1991) Holy Father, "Centesimus annus", encyclical letter to his venerable brother bishops in the episcopate, the priests and deacons, families of men and women religious, all the Christian faithful and to all men and women of good will on the hundredth anniversary of Rerum novarum. http://w2.vatican.va/content/john-paul-ii/en/encyclicals/documents/hf_jp-ii_enc_01051991_centesimus-annus.html. Retrieved 30 November 2016
John Paul II (1995) Evangelium vitae, to the bishops, priests and deacons, men and women religious, lay faithful and all people of good will on the value and inviolability of human life. http://w2.vatican.va/content/john-paul-ii/en/encyclicals/documents/hf_jp-ii_enc_25031995_evangelium-vitae.html. Retrieved 30 November 2016
John Paul II (1999) Address at the Liturgy of the Word, Zamosc, 12 June 1999. https://w2.vatican.va/content/john-paul-ii/en/travels/1999/documents/hf_jp-ii_hom_12061999_zamosc.html. Retrieved 30 November 2016
John Paul II (2001) L'engagement pour éviter une crise écologique majeure, Audience générale, 17 janvier. https://w2.vatican.va/content/john-paul-ii/fr/audiences/2001/documents/hf_jp-ii_aud_20010117.html. Retrieved 30 November 2016
Lyons JA (1982) The cosmic Christ in Origen and Teilhard de Chardin, a comparative study. Oxford University Press, Oxford
Maloney GA (1968) The cosmic Christ from Paul to Teilhard, a comparative study. Sheed and Ward, New York
Mannheim K (1954) Ideology and Utopia: an introduction to the sociology of knowledge. New-York, Harcourt-Brace
Moltmann J (2004) Le rire de l'univers, Traité de christianisme écologique, Anthologie réalisée et présentée par Jean Bastaire. Éditions du Cerf, Paris, p 146
Nys D (1908) Cosmology. In: The Catholic encyclopedia. Robert Appleton Company, New York. Retrieved August 5, 2016 from New Advent: http://www.newadvent.org/cathen/04413a.htm
Paul VI (1967) Populorum Progressio, encyclical of Pope Paul VI on the development of the peoples. http://w2.vatican.va/content/paul-vi/en/encyclicals/documents/hf_p-vi_enc_26031967_populorum.html. Retrieved 30 November 2016
Paul VI (1970) Visit of Pope Paul VI to the Fao on the 25th anniversary of its institution, Monday, 16 November. https://w2.vatican.va/content/paul-vi/en/speeches/1970/documents/hf_p-vi_spe_19701116_xxv-istituzione-fao.html. Retrieved 30 November 2016
Paul VI (1972) Message of his holiness Paul VI to Mr. Maurice F. Strong, Secretary-General of the Conference on the Environment. https://w2.vatican.va/content/paul-vi/en/messages/pont-messages/documents/hf_p-vi_mess_19720605_conferenza-ambiente.html. Retrieved on 30 November 2016

Pontifical Council for Culture, Pontifical Council for Interreligious Dialogue (2003a) Jesus Christ, the bearer of the water of life, A Christian reflection on the "New Age". http://www.vatican.va/roman_curia/pontifical_councils/interelg/documents/rc_pc_interelg_doc_20030203_new-age_en.html. Retrieved 30 November 2016

Pontifical Council for Culture, Pontifical Council for Interreligious Dialogue (2003b) Presentations of Holy See's document on "New Age". http://www.vatican.va/roman_curia/pontifical_councils/interelg/documents/rc_pc_interelg_doc_20030203_press-conf-new-age_en.html. Retrieved 30 November 2016

Pope Francis's Ethics for Democratic Capitalism and the Common Good

Peter Venton

1 The Encyclical Letter, Praise Be to You *Laudato Sì*

In his encyclical letter on the environment, Pope Francis urgently appealed for a new dialogue with all people, on the issue of how we are shaping the future of our planet (Francis 2015, paras. 3, 14). The focus of his contemplated dialogue is the need for an "economic ecology" capable of appealing to a broader vision of reality that integrates economics, politics and humanism for the common good (Francis 2015, paras. 141, 157). According to Francis the common good includes social peace, distributive justice, the eradication of poverty, access to steady employment and, above all, stewardship of the earth's environment (Francis 2015, paras. 157, 127).

The encyclical constitutes an attack on the present variant of capitalism evident in democratic countries throughout the world. In Francis's view, "Politics and the economy have made mistakes and they tend to blame each other when it comes to poverty and environmental degradation. It is to be hoped that they can acknowledge their own mistakes and find forms of interaction directed to the common good (Francis 2015, para. 198)." He especially criticizes the failure to integrate the two silos of economics and politics stating that "economics without politics cannot be justified since this would make it impossible to favor other ways of handling the various aspects of the present crisis" (Francis 2015, para. 196). Furthermore, he maintains that "[p]olitics must not be subject to the economy, nor should the economy be subject to the dictates of an efficiency-driven paradigm of technocracy" (Francis 2015, para. 189). "Today, in view of the common good, there is an urgent need for politics and economics to enter into a frank dialogue in the service of life, especially human life (Francis 2015, para. 189)." For human life, moral

P. Venton (✉)
Ministry of Finance in the Government of Ontario, Toronto, ON, Canada
e-mail: peter.venton@bell.net

L. Westra et al. (eds.), *The Role of Integrity in the Governance of the Commons*,
DOI 10.1007/978-3-319-54392-5_17

renewal is needed to counteract the "ethical and cultural decline" that has featured "rampant individualism" and a "self-centred culture of instant gratification" (Francis 2015, para. 162). This decline has left a citizenry of isolated individuals with a "utilitarian mindset" and an "unethical consumerism" that are "bereft of social or ecological awareness" (Francis 2015, para. 219).

2 Introduction

The main argument in this chapter is that the encyclical criticisms represent an attack on neoliberal capitalism as opposed to democratic capitalism. The argument is advanced in Sect. 3 which defines democratic capitalism and neoliberal capitalism with reference to the United States of America (US). Specifically it is argued that the US came close to exhibiting democratic capitalism in the middle of the twentieth century but that it evolved into neoliberal capitalism in the 1980s and neoliberal capitalism has prevailed to the present day.

In his encyclical, Francis asserts that politics and the economy have made mistakes when it comes to poverty and environmental degradation. Section 4 identifies three mistakes of the US government over the last 80 years and one mistake in the capitalist economic system.

In Sect. 5 the working of capitalist economic systems is analysed to show the inherent conflict between the interests of capitalist firms and the interests of the majority of the public as reflected in the elements of the common good. In a democracy the fundamental role of the government is to mediate the interests of the capitalist firms with the interests of the majority. In democratic capitalism the mediation is balanced in favour of the common good. In neoliberal capitalism the mediation is balanced in favour of the interests of capitalist firms. The key factor behind the conflict is the capitalist firm's adoption of technology for maximizing profits. This adoption has the pernicious effects of reducing access to steady employment and excessive degradation of the environment which are important elements in Francis's definition of the common good.

In Sect. 6 the moral cultural values of democratic capitalism are discussed and listed under four categories. The first three, which pertain to democracy, are responsible citizenship, ethics for citizens in democracy and values that underlie the principles of equality, equal opportunity and social justice. The fourth category comprises values that relate to the efficiency and effectiveness of the capitalist economic system. This section concludes that most of these values are found throughout the encyclical with an important omission which is the virtue of citizens' voting in elections. The significance of this omission for the US is explained in Sect. 7 on Conclusions.

3 From Democratic Capitalism to Neoliberal Capitalism in the US

Democratic capitalism is an ideology that involves a combination of a democratic political system with a capitalist economic system that is based on a liberal moral-cultural system (Novak 1991, p. 14). It includes the principles of freedom and equality and rule by the majority (Simpson 2011, p. 72). "Democratic government is the overarching mechanism for resolving social priorities in the face of market failures (Coyle 2011, p. 251)." In this sense the capitalist economic system is subject to the democratic system. In his book *The Spirit of Democratic Capitalism*, American Catholic philosopher Michael Novak states that democratic capitalism is about three dynamic systems converging as one: a democratic polity, a capitalist economic system based on markets and incentives and a moral-cultural system which is pluralistic and, in the largest sense, liberal. Democratic capitalism is not merely a free enterprise system. Its political system has many legitimate roles to play in economic life, from protecting the soundness of the currency to regulating international trade and internal competition. Its moral-cultural system has many legitimate and indispensable roles to play in economic life, from encouraging self-restraint, hard work, discipline and sacrifice for the future to insisting upon generosity, compassion, integrity and concern for the common good (Novak 1991, pp. 14, 57, 58).

Democratic capitalism, as defined above, prevailed in the US in the 1950s and 1960s in the sense that the capitalist economic system was controlled by the then democratic political system that was supported by the majority and in the sense that there was a high degree of freedom and equality in the country. In his book, *Fractured Republic*, Yuval Levin notes that "[t]he (US) economy of the 1950s and 1960s was exceedingly regulated and constrained" (Levin 2016, p. 51). Further he stated that, "low income inequality was a function precisely of restraints on dynamism" of the economy as a "global economic leader" (Levin 2016, pp. 53, 51). "A highly progressive wartime system of taxation, with marginal income-tax rates topping out at an astonishing 91 percent remained in place until the early 1960s (Levin 2016, p. 52)." "For conservatives the era epitomized exceptional cultural stability and cohesion – what seemed to be a broad and traditionalist moral consensus but also fairly broadly shared prosperity (Levin 2016, p. 53)." The foregoing arrangements could be said to be supported by the majority as the post-war decades involved a "highly unusual elite consensus on a broad range of issues in American politics" (Levin 2016, p. 53). "Even in the early 1960s an overwhelming consensus on many key issues stretched across party lines (Levin 2016, p. 54)." Most of Levin's book is about the fracturing of this mid twentieth century consensus that "grew from diffusion into polarization – of political views, of income, of family patterns and ways of life" (Levin 2016, p. 3).

The world economic context for Levin's description of US society in the 1950s and 1960s is outlined by Geoff Mann:

> The quarter century or so following World War II is often called capitalism's "golden age" or the Long Boom – an era during which capitalist global North (western and northern Europe, North America, and Australia and New Zealand) experienced unprecedented economic growth, low unemployment, increased average living standards, decreasing income and wealth inequality, and a vast expansion of what we now call the welfare state. The following fifteen years or so, however, roughly 1967-82, saw the whole thing seemingly go to pot. These years inaugurated a process we might call the Long Downturn, a trajectory which, depending upon one's data and interpretation, continues today (Mann 2013, pp. 113, 114).

Several forces had come together in the 1960s just before the long downturn that put downward pressure on US corporate profit rates. By the late 1950s US society had reached a stage of unprecedented affluence that moderated the need for increased material consumption. Following the end of the baby boom in 1964, the rate of growth in population declined. Post-war immigration from Europe also slowed. As the 1970s approached international competition resumed as Europe and Japan had rebuilt the industrial structures that had been damaged in the world war. At the same time labour unions retained strong bargaining power that arose in the 1950s due to general labour shortages following the war and reductions in the hours of the standard work week. Increasing participation of women in the labour force did not begin until the early 1970s.

In response to these pressures the US business community started to organize in 1971 for the purpose of countering the then prevailing public criticism of the free enterprise system and to seek new opportunities for maintaining profit rates (Krugman 2009, pp. 118–120). What emerged was neoliberalism which involves policies for reducing taxes on business income, reducing regulations on their activities, increasing government subsidies to business, privatizing government operations, and free trade and foreign investment liberalization. All of these were designed to enhance their profitability either by increasing demand for their products or reducing their costs of operation. The advocated government tax reductions and subsidies of course required government fiscal austerity and, by implication, a contraction in the relative size of the government sector in the economy (Fodor 2013, pp. 105–111).

In 1981 neoliberalism gained political support in the Reagan Administration. That support continued during the 1990s under the Clinton and Bush administrations although, in the latter case, neoliberal strategy appeared to be offset by massive expenditures to fight a war in Iraq. The financial crisis of 2008 prompted a brief return to major expansion in government expenditure by the Obama administration in order to avert another financial collapse and depression of the kind experienced in the 1930s. Since then, there has been no apparent political will in Congress to change the neoliberal course. Specifically there has been no will to raise taxes to fund government investment in infrastructure that would generate more employment. During his election campaign Donald Trump promised to reduce the tax rate on corporate income.

3.1 *Criticisms of Neoliberal Capitalism in Laudato Sì*

In two respects the criticisms in the papal encyclical constitute an attack on the main features of neoliberal capitalism. First, the US government's adoption of neoliberal policies allowed the economic system to dominate the political system. Francis's criticism that "politics must not be subject to economics" is, in effect, a criticism of neoliberal capitalism which features an economic system that dominates the political system (Francis 2015, para. 189). The criticism implicitly supports one particular feature of democratic capitalism, namely that businesses in the economy are subject to the political system.

Secondly, the domination of the economic system over the political system in neoliberal capitalism severely reduces the alternatives available to the government for addressing issues such as low economic growth, unemployment, growing income inequality or climate change. Raising taxes and regulations are not options. In neoliberal capitalism the only policy instruments left for the government are the monetary policy of the central bank (federal reserve) and political "moral suasion" that political leaders sometimes employ to encourage changes in public opinion or business behaviour. Francis's critique that "economics without politics cannot be justified since this would make it possible to favour other ways of handling the various aspects of the present crisis" is a criticism of neoliberal capitalism. It implicitly supports the feature of democratic capitalism that leaves all possible alternative policy instruments open to the government (Francis 2015, para. 196).

4 Mistakes of Politics and the Economy

As noted earlier, Francis "hopes that politics and the economy can acknowledge their own mistakes and find forms of interaction directed to the common good" (Francis 2015, para. 198). Here, politics appears to relate to the political regime and economy appears to relate to the economic system. In democratic capitalism the role of the government is to resolve the social priorities of the citizenry that define the common good and to correct market failures in the economic system. This section describes three mistakes that the US government made as it evolved from democratic capitalism in mid twentieth century to neoliberal capitalism in the latter part of that century and one mistake that is inherent in capitalist economic systems.

4.1 *Failure to Heed the Advice of Economists*

Over the past 80 years the US government failed to heed the advice of prominent economists who concerned themselves with the relationships between political systems and the detailed workings of capitalist economic systems.

Foremost among the economists was John Maynard Keynes "whose influence towers over the study of modern economics (Stiglitz 2010, p. xii)." For many years Keynes had observed the phenomenon of chronic unemployment in capitalist market systems. In the 1930s his revolutionary prescriptions were to modify the market economy through progressive taxation to affect income and wealth distribution and to socialize investment to achieve full employment (Minsky 2008, pp. 146, 147). Specifically Keynes's advice was necessary to prevent the tendency of the economic system toward increasing inequality in wealth and income that leads to unemployment. As a consequence some would not have access to steady employment which is an element in Francis's definition of the common good. Otherwise the economy would tend to ever increasing inequality of income and wealth, some of which would not be recycled back into the real economy but remain "bottled up" in the financial sector and not spent. In simple terms the main reason for this is that the wealthy, having more income than they need, would put their savings in the bank or invest them in other financial assets such as bonds and stocks. They might invest their savings in existing assets such as real estate, or art or other collectibles but this would drive up the prices of these assets or encourage the middle class to sell what modest assets they had in order to cover their living costs for a time. Eventually some of the middle class would lose all of their modest wealth and become poor, the poor being defined as those having virtually no assets. Over time the rate of economic growth would decline because of insufficient demand for goods and services from the middle class and the poor.

The US Government ignored Keynes's advice and took the opposite tack. It reduced taxes on the profits of business firms with the hope that they would invest it in real capital to produce more output and thereby create jobs. However tax reductions forced the government to reduce its infrastructure expenditures, in part in order to balance its budget—expenditures that would otherwise have increased economic output and employment. This approach has the effect of increasing "corporate welfare" which is the difference between the value of all the benefits that corporations receive from the government every year and the taxes that they pay to the government. These include social investments such as public research, roads, industrial parks and human capital. They also include social expenses for public schools, day care provision, social security and unemployment insurance. And they include expenditures on health for the costs of pollution generated by firms. Tax reductions on corporations need to be recouped by taxes on the middle class and the poor through, for example, sales taxes. Eventually the middle class and the poor resist paying what appear to be excessive taxes.

Almost 70 years after Keynes died, French economist Thomas Piketty confirmed Keynes's hypothesis about the workings of a capitalist economy. On the basis of his analysis of 200 years of data on the dynamics and structure of wealth in capitalist economic systems in France and England, Piketty discovered an empirical phenomenon that the rate of return on capital always exceeded the rate of growth in the national income (Piketty 2014, p. 354). The mathematical consequence of that fact is that wealth always grows faster than national income. Consequently, after a few generations, inherited wealth begins to dominate the economy as wealth eventually becomes very concentrated in the hands of a few. A related manifestation of the trend is that the share of income from the hard work of labour in the economy shrinks while the share of easy income from capital (i.e., wealth) increases. For clarification, income from capital includes rental income from real estate, imputed rental income from owner occupied residences, dividends and capital gains from stocks, interest from savings accounts in banks and bonds. It is the shrinking share of labour income in the economy that presents a problem for the capitalists because it results in insufficient consumption to meet the output of the productive capital of the capitalist firms with the result of an under utilization of capital and a falling rate of profit. In these circumstances, "capitalists dig their own grave: either they tear each other apart in a desperate attempt to combat the falling rate of profit . . . or they force labor to accept a smaller and smaller share of national income, which ultimately leads to a proletarian revolution and general expropriation" (Piketty 2014, pp. 228, 229).

In the US signs of discontent about income inequality emerged in the Occupy Wall Street movement in 2011. In that year it was revealed that the share of total income earned by the top 1% of income earners had increased from 10% in the 1950s to 25% in 2010—a share not seen since the 1920s just before the stock market crash of 1929 that ushered in the great depression of the 1930s (McQuaig and Brooks 2011, p. 32; Stiglitz 2011). Signs of discontent of the proletariat surfaced in the US Presidential campaign of 2015 and 2016.

In 240 years of US history, as in Britain and France, the only time when the rate of return on capital came close to the growth rate in national income was after World War 11. This occurred because of special circumstances of huge markets for US goods owing to the destruction of the industrial structure of its competitors in Europe and Japan in World War II and also the very high tax rates on corporations and high income earners in a democratic capitalism regime.

4.2 Adoption of a Narrow Focus on Economic Growth

In the 1950s the US government (as well as Canadian and Australian governments) adopted economic growth as a paramount goal—a policy that continued into the 1970s despite the reservations of several prominent economists such as John Kenneth Galbraith, Kenneth Boulding, Ezra Mishan and Herman Daly (Victor 2008, pp. 14–15). Their general criticism was that the focus on economic growth ignored all of the other elements of the common good.

In the 1980s a General Progress Indicator (GPI) was developed for the US—a comprehensive set of 25 indices of elements of the common good. It recognizes the interdependence of economic well-being, the quality of the natural environment and the quality of social relationships. In the economic domain it measures changes in personal consumption expenditure adjusted for changes in income inequality, consumer durable service flows, services from public infrastructure, net capital investment and net foreign borrowing. In the environmental domain it assigns costs to air, noise and water pollution, lost farmland, lost wetland and lost forests, depletion of oil reserves and carbon dioxide and ozone damage. In the social domain it counts the benefits of volunteer work, higher education, and parenting as well as the costs of crime, inequality, commuting and auto accidents. (Talbeth et al. 2007, pp. 4, 1). Over the 34 year period between 1970 and 2004, the US GPI grew by only 5% while real per capital GDP increased by 105% (Victor 2008, p. 129 Chart). It reveals how little overall progress in US society has been generated by economic growth.

The wave of protests associated with the Occupy Wall Street movement in 2011 served to dramatically highlight the burgeoning awareness that economic growth would not generate the vast array of benefits claimed for it. Those protests sought in part, at last, to expose the false hope created by the neoliberal proponents of economic growth. One of the eight themes in the encyclical that reappears throughout the encyclical is "the call to seek new ways of understanding the economy and progress" (Francis 2015, p. 16). This call is a logical response to the mistaken assumption that social progress is highly correlated with economic growth.

4.3 A Lost Opportunity to Limit Economic Growth in Mid Twentieth Century

In retrospect a major opportunity to limit US economic growth arose in the mid twentieth century when John Kenneth Galbraith published his 1958 book, *The Affluent Society*. Galbraith claimed that American households had reached a state of affluence that more than met the Aristotelian ideal of having enough income for sustenance and leisure for politics and philosophy (Newell 2011, p. 17). His advice to the American people and their government was to spend more on social security to reduce social tensions and less on private consumption and focus on enhancing the quality of life (Parker 2006, p. 289). Galbraith's advice went unheeded. Instead, "[i]n a Cold War state the government carried out its Keynesian macromanagerial mandate under circumstances heavily determined by military and strategic goals; meanwhile, the domestic economy was left to advertisers and marketers in an endless celebration of private consumption. This consumption increasingly met not needs so much as wants carefully manipulated by private-goods producers. Pressing public needs were not filled (Parker 2006, p. 302)."

In an October 13, 1954 speech at Wellesley College Massachusetts on "Economic Freedom", Galbraith argued that:

> The achievement of democracy and market capitalism was that both created the preconditions for the exercise of freedom, but at the same time market capitalism generated its own powerful limitations on freedom promoting the coarse ideological claim that ever-increasing consumption is freedom. To Galbraith this was a "false ideal". He argued, or hoped, that, given more affluence, more and more people would come to realize that "leisure, free time and intellectual achievement are the real thing". (Parker 2006, pp. 302, 303, 724n89)

The outcomes of the US government's mistaken focus on economic growth are a major depletion in its natural resources and huge deficits in public infrastructure. GPI measures indicate that, between 1960 and 2004, resource depletion increased by 506% (Talbeth et al. 2007, pp. 21, 22, Table 1). As of 2013, the US needs an estimated $3.6 trillion dollars to restore its infrastructure according to the American Society of Civil Engineers 2013 Report Card (ASCE 2013).

4.4 A Mistake in Capitalist Economic Systems

As will be explained later in Sect. 6.3.1, nineteenth century neoclassical economists made a mistake by insisting that wage rates be determined in labour markets rather than on the basis of the labour theory of value that classical economists had used. The results have been wage rates below poverty lines for decades—an outcome entirely inconsistent with the democratic principle of equality. Consequently the freedom from work has been dramatically reduced for a significant proportion of US workers who need to work at more than one job to make ends meet. The US government did fix this problem by legislating minimum wages in the late 1960s that were above the poverty line for a family of three. However, since then the government has allowed the purchasing power of the minimum wage to decline by 30% over the period from 1968 to 2012. (US Department of Labor 2012, Chart 1). In 2013 it was almost 20% below the poverty line for a family of three (Cooper 2013).

5 The Nature of Capitalist Economic Systems

Capitalist economic systems comprise four institutions: private firms for producing and distributing goods and services, market exchange, a monetary system based on the production of bank-credit money and a distinctive role of the state in relation to these three foregoing institutions (Mann 2013, p. 13).

The reality of the relationship between firms and the government is one of mutual dependence that is described by Canadian political scientist Geoffrey Hale as "an uneasy partnership" (Hale 2009, paras. 7–32). The government supports the private firms with "social investments" that enhance their productivity, by incurring "social expenses" that reduce their labour costs and by Keynesian policies that periodically increase aggregate demand for their products. "Social investments" include for example expenditure for public research, roads, industrial parks, and human capital. "Social expenses" include for example expenditures for public schools, day care provision, social security, and unemployment insurance" (Lichty 2005, pp. 1–5). In return it receives tax revenues to fund operations. The positive difference between the costs of the above government services and the tax revenues from the corporations represents corporate welfare which is a hallmark of neoliberal capitalism.

The partnership between the government and private firms is uneasy because there are inherent conflicts between the mission and interests of private firms and those of the government in representing the citizenry for the common good. Private firms are organized to enhance the production of goods and services and to facilitate the distribution and sale of goods and services in exchange for money for the purpose of maximizing profits. Firms use human labour in combination with real capital (e.g. tools, machinery, equipment, factories) that is either made by the firm or purchased from other firms to produce the product. It is the firm's use of capital in combination with labour that gives rise to the name, capitalism.

A key factor for the firm's success in maximizing profits for the shareholders/ owners is size in terms of sales and production. As sales and production increase, some costs are fixed with the result that sales increase faster than costs and profits increase even faster. As the level of sales and production increases there are more opportunities to specialize in certain areas of production—specialization that increases productivity which reduces costs per unit of output and increases profits. Furthermore, as sales and production increase, the firm is able to crowd out competing firms and dominate the industry or market sector in which it operates. As a result the firm will have greater market power. This "monopolistic" power enables it to set higher prices for its product with little risk that competitor firms will respond by reducing their prices to take sales away. Its "monopolistic" power also enables it to determine its rate of profit through a combination of setting prices and restricting its investment in productive capital. In summary the mission of private firms is for unlimited expansion and their ultimate success is achieved by dominating the industry or market in which they operate. Their success is limited only by the size of the market.

Opportunities for specialization and enhanced profits that are enabled by the larger size of the firm are intricately bound up with the adoption of technology. Today, the key factor for the accumulation of profits and wealth is technology. Technology enables the substitution of non-human capital for human labour in the production process. Increasing the investment in non-human capital increases the size of the annual return of the costs of non-human capital which is roughly equal to the amortization of the cost of that capital plus the interest on funds borrowed to acquire it. This is a first claim on the annual revenues of the private firm in the sense that it is a return even if there is no annual profit of the firm. The larger effect of investment in nonhuman capital that embodies technology is increased profits that result from the reduction in the quantity and/or cost of labour in the process of production and distribution. Given the typically high ratio of labour costs to sales revenue (e.g., 50%) in private firms relative to the low ratio of profits to sales revenue (10%), a given percentage reduction in labour costs results in a much larger percentage increase in profits. The mathematics of this substitution represents a powerful incentive for investment based on new technology.

Empirical analysis by economist Robert Solow concluded that most of economic growth in capitalist economies is the result of "technological factors". Solow won a

Nobel Prize for his 1956 growth theory. His study incorporated an econometric model that showed that only 10% of the variation in economic output in a nation's economy was due to the factors of labour and capital. The other 90% was due to something else which he called 'technical change' (Mazzucato 2013, p. 33).

The conclusion of Solow's study explains Francis's statement that "[t]he technocratic paradigm also tends to dominate economic and political life" (Francis 2015, para. 109). It also illustrates Francis's point that "technocracy coexists with the ... extreme which sees no special value in human beings" (Francis 2015, para. 118). As well it explains his argument that "profit cannot be the sole criterion to be taken into account" in considering technological innovations (Francis 2015, para. 187). Finally it explains why technology impacts negatively on access to steady employment, a goal which Francis argues "must be prioritized no matter the limited interests of business and dubious economic reasoning" (Francis 2015, para. 127). In this light the only way technology can be rationalized is by somehow increasing demand and consumption for other new products that will require some labour to produce. Economic growth, regardless of its utility or intrinsic value, is imperative for the capitalist firms. But the unregulated outcomes of the operation of private firms and technology run contrary to the achievement of the several elements of the common good.

The role of the government in the capitalist economic system is to mediate the conflicting interests of the private firms and the interests of the majority of citizens in the common good. The outcome of that mediation depends on the political power of the citizenry relative to the power of the capitalist firms. In democratic capitalism, where the political system is relatively strong, the outcome will favour the common good. In neoliberal capitalism where the political system is relatively weak, the outcome will favour the firms in the economic system over the common good.

6 Moral Cultural Values of Democratic Capitalism

The moral cultural values of democratic capitalism are grouped into four categories: responsible citizenship, ethics for citizens in democracies, values underlying democratic principles and personal values related to the economic system.

6.1 Responsible Citizenship

These values include voting in elections, engaging in public discourse about social, political and economic issues and elements of the common good. They include learning about civics, the theory of democracy and other political regimes, political science, principles and precepts of public economics, and the sociology of social movements. In general, learning is about aspects of philosophy, religion, history, and language that are relevant to critical and imaginative thinking for the assessment of, and solutions to, social, political and economic issues.

6.2 *Ethics for Citizens in Democracy*

These values pertain to citizens who are law abiding, who pay taxes, who are willing to sacrifice some freedom for the common good and who have moral sentiments of empathy, common sympathy, and benevolence (in terms of charity and community service) and who have a propensity to consider the points of view of others.

6.3 *Values Underlying Democratic Principles*

Three main principles of democracy are the objectives of equality, equal opportunity and social justice that are capable of measurement. However, the objectives may be associated with values which are abstract or subjective and hence not quantitatively measurable.

6.3.1 Equality

Of course, the term equality is a mathematical one that implies measurements of two or more things or conditions. A problem with liberal democracy has been the absence of definitions of the respect or respects in which humans may be considered to be equal. Nineteenth century classical economists utilized the labour theory of value to define conditions of equality. The labour theory of value holds that the value of commodities and things is proportional to the amount of labour that goes into their production. Further they considered the value of labour of all kinds in a community to be akin to some average that would vary by time and place (Mann 2013, pp. 37, 38). For a given time and place, all labourers could be considered equal if their hourly rate of pay was at the community average or at least at some standard value. In this way they linked economic precepts with the democratic principle of equality. The abstract value associated with equality could, for example, have been dignity on the premise of human psychology that those who are perceived to be equal to others may be more highly regarded than they would be if they were perceived not to be equal. Democracy is also associated with notions of fraternity which implies equality of status.

Nineteenth century neoclassical economists made a mistake when they decided that wages for labour should be determined in markets for labour, like markets for commodities. However, the reality is that the two markets are different in the sense that they have different objectives. For commodities markets are indispensable for balancing the demand with supply where there are many individual buyers in the market. Prices in these markets send signals to both buyers and sellers and suppliers respond by making adjustments to prices which affect the decisions of potential buyers that result in the clearing of the market. Economist Thomas Piketty explains that "[t]he price system (in markets) plays a key role in coordinating activities of millions of individuals – indeed, today, billions of individuals in the new global economy. The problem is that the price system knows neither limits nor morality (Piketty 2014, p. 6)." The moral implication of this statement is that markets should not apply to human labour. The labour theory of value serves a different objective than clearing the market; its objective is equality.

6.3.2 Equal Opportunity

The objective of equal opportunity relates to Aristotle's views of economics in democratic regimes. For Aristotle "the primary purpose of household management is to give citizen-householders the leisure for pursuing man's two chief ends, 'politics and philosophy" (Newell 2011, p. 17). All citizens should therefore have an equal opportunity to participate in the political and social life of society. To this end all need at least a threshold income sufficient to provide for the basic necessities of life (e.g., food, shelter, clothing) plus an extra amount to enable them to engage in political and social life. In modern terms threshold income can be established on the basis of a standard work week multiplied by a minimum hourly rate of pay for all labour. An underlying value of this condition is that, in these circumstances, all labouring individuals could be seen to have an equal amount of freedom from work (Koivukoski 2015, p. 174). Freedom is a higher value associated with the objective of equal opportunity. More precisely the value is equal freedom for all.

One respect in which all citizens are objectively roughly equal is the amount of time that they have available for work over the course of their lives—given their biological needs for sleep, the cultural norms of society on the standards for work, vacations and homemaking. For example in a society, the hours available for work may range from a standard of 35 h to 70 h which may be a maximum considered to be feasible for most. This example implies a range of pay illustrated by an index from 100 (at 35 h) to 200 (at 70 h) per week. The resulting range of income would be considerably smaller than that which prevails today in democratic nations with advanced economies such as Canada and the US.

By contrast wage rates that result from the operation of the laws of supply and demand in labour markets generate hourly rates well below the threshold for some. Consequently they might induce individuals to take more than one job with the consequence that their freedom from work would be greatly reduced. In this circumstance they would become in effect "wages slaves"—a circumstance that would violate Aristotle's notion of the second kind of freedom which is "to do as one likes because not to do as one likes is the life of a slave" (Cooper 2011, p. 200).

The labour theory of value can be extended to those with higher skills by adjusting it to account for the time that they must expend to acquire the skills.

Specifically the threshold labour theory of value amount per hour would be increased to compensate them for the potential years of work that they lost in order to study to acquire the skills. For example, the rate of pay for a medical doctor who had spent 10 years of study from age 18 to 27 in order to practise medicine for 38 years from 28 to 65 would be higher than another person who did not undertake the same level of study and therefore could work for 48 years from age 18 to 65. Specifically the doctor's hourly rate of pay would be set 26% higher—a figure derived from the ratio of 48 years of work to 38 years of work minus 1. Consequently, at the end of his or her working career the cumulative pay of the doctor over his or her career would be equal to the cumulative pay of the other person (who did not study) over his or her longer working career.

The criticism of the application of the labour theory of value is that there is no economic incentive for workers to change jobs that are in scarce supply. So pay rates based on demand and supply would need to apply in circumstances where there was a general shortage of labour at the threshold wage rates. Unemployment arising from higher labour-theory-of-value wage would have to be absorbed into the government sector for providing needed government services or retraining for other opportunities in the private sector of the economy.

The foregoing does not apply in the many institutions in today's society where the laws of supply and demand are suspended and replaced with "administered rates" of pay that are based upon rules of thumb, labour laws or other factors. Examples of these institutions are government, schools, universities, hospitals etc. ... in the so called broader public sector, labour unions in the private sector and chief executive officers (CEOs) of private firms. In the case of the latter compensation is based on historic cultural norms and the bargaining power of CEOs vis a vis representatives of company shareholders.

6.3.3 Social Justice

Social justice is about ideas of what constitutes tolerable inequality among citizens. The work of John Rawls is instructive on the issue of social justice (Rawls 1999). He developed the idea of a contractual theory of justice which differs from the classical utilitarian principle of justice. A utilitarian contract would be considered ideal if it achieved the greatest net balance of satisfaction summed over all individuals who agreed to the contract—even though some would suffer a net loss in satisfaction. By contrast a fair contract is one in which no one would suffer a loss and many would realize some positive satisfaction (Rawls 1999, p. 20). A contractual theory of justice appears to be particularly suited to democracy because a central feature of democracy is the implicit social contract wherein citizens voluntarily give up some of their individual liberty in exchange for collective benefits (Venton 2015, p. 206).

6.4 *Personal Values Related to the Economic System*

These include hard work, self restraint, discipline and sacrifice for the future. Personal responsibility for living within one's means are assumed to be an important virtue; without them individuals become burdens on the state and by implication a burden on their fellow citizens.

6.5 *Comparisons of Moral Cultural Values with the Encyclical*

The application of the labour theory of value to define and measure the democratic objectives of equality and equal opportunity implicitly treats humans as subjects as opposed to objects such as commodities. In this respect they are consistent with Francis's interpretation of the biblical accounts that "invite us to see each human being as a subject who can never be reduced to an object" (Francis 2015, para. 81).

The value of fairness incorporated in social justice in conjunction with the values of freedom underlying equality and equal opportunity would eliminate poverty and address Francis's criticism that "[w]e fail to see that some are mired in desperate and degrading poverty, with no way out, while others have not the faintest idea of what to do with their possessions, vainly showing off their supposed superiority and leaving behind them so much waste which, if it were the case everywhere, would destroy the planet" (Francis 2015, para. 90).

Francis agrees with the importance of citizen engagement in public discourse as evidenced in his statements that "people who profess to be believers (in a religion) . . . should dialogue among themselves for the sake of protecting nature, defending the poor and building networks of respect and fraternity" (Francis 2015, para. 201). "The gravity of the ecological crisis demands that we all look to the common good, embarking on a path of dialogue which demands patience, self discipline and generosity, always keeping in mind that realities are greater than ideas (Francis 2015, para. 201)."

In summary most of the moral cultural values of democratic capitalism are found in the encyclical. An important exception is the virtue of voting in elections. The significance of this omission for the US is outlined in the concluding section.

7 Conclusions

Most of the values in the moral cultural system of democratic capitalism are found in the encyclical and this fact supports the implicit argument in the title of this chapter that Pope Francis's ethics serve democratic capitalism. However, a notable omission is a reference to the virtue of voting in elections. In the US this is a significant omission because low voter turnout tends to support neoliberal capitalism rather than democratic capitalism.

In the 2004 introduction to the second edition of his 1991 book, *Why Americans Hate Politics*, American syndicated columnist E.J. Dionne Jr. makes the following observation:

> The politics of the 1960s shifted the balance of power within the liberal coalition away from working class and lower middle class voters whose main concerns were economic. What is striking about political events in the 1960s was that they allowed both the nations' dominant ideologies and parties to become vehicles for upper middle class interests. … While upper middle class reformers, left and right argued about morality, anti communism, imperialism and abstract rights, millions of working-class and lower middle class voters were confined to the sidelines wondering why the nation's political discussion had become so distant from their main concerns which were economic. (Dionne 2004, p. 13).

The reality appears to be that the upper middle class are likely to constitute a large segment of the voting electorate. For example, in the November 2010 US Congressional elections, 60% of voters earned income of more than $50,000 despite the fact that they represented only 40% of the population. This reflects the fact that 68% of the population over the age of 18 that earned more than $50,000 voted, while only 30% of the population over the age 18 who earned less than $50,000 voted. (Krugman 2009, p. 193; Venton 2015, p. 211).

At the same time the US tax and expenditure system is such that taxes are roughly proportional to the earned income of citizens while government expenditure benefits are roughly proportional to population. Consequently, it is the case that taxes on the affluent exceed their expenditure benefits whereas the very opposite applies to the less affluent except for a very small number of the very poor. In these circumstances the majority of voters are affluent and they have a rational preference for lower taxes and government austerity. In other words the US political system could be characterized as government of the affluent by the affluent for the affluent. In the absence of a voting public who are willing to sacrifice some freedom for the common good, democratic capitalism requires compulsory voting and an education system focused on responsible citizenship for enlightenment on politics and public economics.

References

ASCE (2013) About America's infrastructure. American Society of Civil Engineers 2013 Report Card. http://www.infrastructure.org. Accessed 13 Nov 2016

Cooper C (2011) Oligarchy and the rule of law. In: Tabachnick D, Koivukoski T (eds) On oligarchy: ancient lessons for global politics. University of Toronto Press, Toronto, pp 196–216

Cooper D (2013) The minimum wage used to be enough to keep workers out of poverty–it's not anymore. Economic Policy Institute, Washington (4 Dec). Available at www.epi.org/publications/minimu-wage-workers-poverty-anymore-raising. Accessed 23 Nov 2016

Coyle D (2011) The economics of enough: how to run the economy as if the future matters. Princeton University Press, Princeton

Dionne E (2004) Why Americans hate politics. Simon and Schuster, New York
Fodor M (2013) Taxation and the neo-liberal counter-revolution: the Canadian case. In: Himelfarb A, Himelfarb J (eds) Tax is not a four letter word: a different take on taxes in Canada. Wilfrid Laurier University Press, Waterloo, pp 101–117
Francis P (2015) Praise be to you Laudato Si. Ignatius Press, San Francisco
Hale G (2009) Uneasy partnership: the politics of business and government in Canada. University of Toronto Press, Toronto
Koivukoski T (2015) Seeking peace in nature: a reading of thoreau on ecology and economy. In: Koivukoski T, Tabachnick D (eds) The question of peace in modern political thought. Wilfrid Laurier University Press, Waterloo, pp 161–182
Krugman P (2009) The conscience of a liberal. W.W. Norton & Company, New York
Levin Y (2016) The fractured republic: renewing America's social contract in the age of individualism. Basic Books, New York
Lichty W (2005) Lecture on James O'Connor's Fiscal Crisis of the State, 1970 4 April. Available at www.umn.eduw-rlichty/Radical%Lectures/radles10.pdf. Accessed 18 Feb 2013
Mann G (2013) Disassembly required: a field guide to actually existing capitalism. AK Press, Oakland
Mazzucato M (2013) The entrepreneurial state: debunking public vs. private sector myths. Anthem Press, New York
McQuaig L, Brooks N (2011) The trouble with billionaires. Penguin Books, Toronto
Minsky H (2008) John Maynard Keynes. McGraw Hill, Toronto New edition of the 1975 book
Newell W (2011) Oligarchy and Oikonomia: Aristotle's ambivalent assessment of private property. In: Tabachnick D, Koivukoski T (eds) On oligarchy: ancient lessons for global politics. University of Toronto Press, Toronto, pp 3–23
Novak M (1991) The spirit of democratic capitalism. Paperback edition with an afterword to the 1982nd edition. Madison Books, New York
Parker R (2006) John Kenneth Galbraith: his life, his politics, his economics, Perennial Edition. Harper Collins, Toronto
Piketty T (2014) Capital in the twenty-first century. Belknap Press of Harvard University Press, Cambridge
Rawls J (1999) A theory of justice. Revised Edition of the 1971 book. Belknap Press of Harvard University, Cambridge
Simpson P (2011) A corruption of oligarchs. In: Tabachnick D, Koivukoski T (eds) On oligarchy: ancient lessons for global politics. University of Toronto Press, Toronto, pp 70–89
Stiglitz J (2010) Freefall: America, free markets and the sinking of the world economy. W.W. Norton & Co, New York
Stiglitz J (2011) Of the 1%, by the 1%, for the 1%. Vanity Fair (May). Vanityfair.com/news/2011/05/top-0ne-percent. Accessed 6 June 2012
Talbeth J, Cobb C, Slattery N (2007) The genuine progress indicator 2006. Redefining Progress Institute, Oakland
US Department of Labor (2012) Minimum Wages U.S. Department of Labor Chart 1, Minimum Wages in 2012 Dollars. dot.gov/featured/minimum-wage/chart1
Venton P (2015) Radical changes in Canadian democracy: for ecology and the 'public good'. In: Westra L, Gray J, Karageorgou V (eds) Ecological systems integrity: governance, law and human rights. Earthscan/Routledge, Abingdon, pp 201–219
Victor P (2008) Managing without growth: slower by design, not disaster. Edward Elgar Publishing, Cheltenham

Natural Catastrophes and Forms of Catastrophism. A New Ethical and Moral Framework Leading Towards the "Responsible Catastrophism Model"

Marco Ettore Grasso

1 Introduction: What Is "Catastrophism"?

The origin of the concept of "catastrophism" derives from the theory of catastrophism, also known as the "theory of cataclysms". It was designed by Cuvier (1827) and explains, without refuting the dogma of creation, the evolution of living beings through sudden and violent upheavals of the earth's crust, after which the living entities, largely destroyed, would have been replaced by other forms of life created again, by virtue of a plan that would end with the appearance of man on Earth.[1] Generally, in the common sense, this notion takes a different meaning, characterized by a negative connotation. In effect, it can be defined as the postmodern tendency to see the catastrophe even where there is not, or the tendency to amplify over time and space the negative effects of an adverse event.

The concept of "catastrophism" inevitably presupposes that of "catastrophe" or "disaster". The term "catastrophe", from greek "*kata*" and "*strepho*", refers to a sudden and dramatic event that disrupts the social life. One connotation of this concept seems to be represented by the character of unpredictability. Nevertheless, we should recognize that in the past a number of natural disasters could have been avoided by humans as predictable.[2] We can therefore distinguish natural disasters from artificial disasters, aware that behind many natural catastrophes often lurks the hand of man.[3] The concept of disaster is a complex concept, because it involves many elements that intersect each other. The natural catastrophe, in fact, is already

[1]This theory was subsequently discussed by other authors, such as Hutton (1785) and Lyell (1830).

[2]An example is the Vajont disaster (1963).

[3]The floods, for instance, are the result of global warming which, as we know, is determined predominantly by human action.

M.E. Grasso (✉)
University of Milan Bicocca, Milan, Italy
e-mail: marco.grasso2@unimib.it; marcoeg@hotmail.it

L. Westra et al. (eds.), *The Role of Integrity in the Governance of the Commons*,
DOI 10.1007/978-3-319-54392-5_18

perceived as a social and moral catastrophe. The link between catastrophes and social conflicts, moreover, has been widely studied in literature (e.g. Nel and Righarts 2008; Slettebak and De Soysa 2010).

This study intends to build a new concept of catastrophism, seen as an ethical and moral approach, which aims to study the "science of disasters", up to reach the "responsible catastrophism model". For this purpose, the paper identifies three branches or study pillars, relative to the ethics of catastrophes.

2 Three Study Areas About Disaster Science: Which "Catastrophism"?

My studies lead me to identify three research pillars or perspectives[4] on the topic of natural disasters, examined from an ethical and social point of view: (1) the branch about human limitations, (2) the study area on the cooperative-preventive character and (3) the solidaristic-transformative view. Each of these dimensions is linked to different meanings of the concept of "catastrophism".

The first pillar concerns human limitation compared to disasters and it is connected to the analysis of the concept of "vulnerability" in social science of disasters. This study area, thus, perceives the "catastrophism" concept in its nature of powerlessness.

The second pillar, instead, refers only to the cooperative-preventive attribute, or rather to the causes of disasters, which are often linked to the selfish actions of man. It follows a concept of catastrophism, which is less pessimistic, "more prudent" and inherent in public policies promoting prevention and risk assessment.

The third pillar, finally, sees the disaster as an opportunity for change, therefore as something positive. Catastrophism here takes on a new "optimistic" perspective.

3 The First Pillar: Awareness of Human Limitations, Disasters and Vulnerability

In my opinion, a first study area about ethical and social science on natural disasters deals with the relationship between human limits and power of nature. This area, which could be defined as the "pillar of the human limit", relates to the question: how man arises in relation to disasters? Weakness, helplessness and awareness of the harmful effects, including deaths, are the main elements characterizing the social and psychic states of the affected communities.

[4]Terms such as "study/research pillars", "branches", "perspectives/dimensions" or "study area/view" are used as synonyms in this work.

Over the past 20 years, 953 disasters have caused nearly 88,671 deaths in Europe. The floods alone have affected more than 5.5 million people (European Commission 2013a). The heat wave that occurred in 2003 determined about 24,000 deaths, causing 13 billion euros loss in European agriculture alone. Estimates suggest that by 2080 millions of people will be experiencing flooding every year due to sea-level rise[5] (IPCC 2007).

According to Voltaire (1756), for example, each of us should reflect on the weakness of our nature. Despite everything being predetermined, or ordered from the hand of Providence, not everything would be arranged in favor of our happiness. For this philosopher, man confronted with catastrophes is unfortunate, the Earth is unhappy and pains are useless, while the worms, which feed on the corpses, become sad comforters of human misery.

Within this branch, therefore, the study of the scientific concept of "vulnerability" is fundamental. The IPCC Third Assessment Report (2001) provides one definition of "vulnerability", considered as a function of exposure, sensitivity and adaptive capacity. Consistently with the IPCC Fifth Assessment Report (2014), on the contrary, vulnerability is defined as the propensity or predisposition to be adversely affected. As specified by Schilling et al. (2012), regardless, the concept of vulnerability lacks one universally accepted definition (Füssel 2007). However, this concept varies considerably in relation to different fields in which it is used. Personally, I identify three meanings of the vulnerability concept applied to the social field of disasters. Indeed, this concept can be considered as: (1) Sensitivity, propension, probability and predisposition to be affected by disasters (Allen 2003; Weichselgartner 2001; Cutter 2003; Birkmann et al. 2014); (2) Exposure to shocks, alarms[6] and stresses about natural hazards (Myers et al. 2008; Tierney 2006); (3) Adaptive capacity, resilience and the ability to cope with the consequences of disasters (Kelly and Adger 2000: 328; Wisner et al. 2003).

I think that the local institutions and their ability to perform an activity mediating between the affected community and the regional and central government levels cover an important role within the concept of social vulnerability linked to natural disasters. This institutional ability, indeed, influences the development of the communities damaged, for example with regard to their ability to reorganize themselves or vice-versa to extend the emergency, becoming increasingly

[5]In the filed of climate change, WHO (2009) estimated that this change had caused about 141,000 excess deaths annually by the year 2004 (in respect with the period 1961–1990), because of disease and malnutrition (WHO 2009). The Global Humanitarian Forum (2009), furthermore, concluded that about 300,000 people die every year because of climate change (for the period 1980–2005). A report by Development Assistance Research Associates and Climate Vulnerable Forum (2012), instead, puts the current annual death toll from climate change at 400,000 (nearly all in the developing world) and projects that by 2030 will rise to nearly 700,000.

[6]Regarding several warning systems, some good practices emerge: the "Indian Ocean Tsunami Warning System", for instance, was established after the 2004 disaster. It now sends alerts to national warning centers within ten minutes of an earthquake. This system has helped to reduce the lives lost in the region. Even so, the development on preparedness systems has generally been limited.

dependent on external aid. Therefore, social vulnerability depends on an amount of socially generated stressors, such as class, livelihood sources, race, religion, ethnicity, gender, technological capacity, socio-political values, inequality, institutional capacity, information deficit, poverty, income distribution and endowments (Ford et al. 2010; Turner et al. 2003; Vincent 2004). As said by Naumann et al. (2014), the vulnerability of a society to disasters depends on several factors such as population, technology, policy, social behavior, land use patterns, water use, economic development and cultural diversity. Vincent (2004) makes a summary of variables, indicators and data sources on this topic and illustrates five determinants of social vulnerability: economic well-being and stability, demographic structure, institutional stability and strength of public infrastructure, global interconnectivity and natural resources dependence. In relation to the last element, Morzaria-Luna et al. (2014) show how in the Northern Gulf of California (Mexico) social vulnerability was higher in communities with higher fishing dependence and lower socioeconomic diversification.

This first pillar, thereby, welcomes the saddest aspect of the "catastrophism" concept, namely, the human and social fragility arising in the face of catastrophic scenarios.

4 The Cooperative-Preventive Pillar

This second area of study relates mainly to the analysis of the causes of natural disasters, as the causes of selfishness and utilitarianism. Indeed, natural disasters often are more harmful in fragile areas where human selfishness has wanted to build houses. Rousseau (1764), in this respect, in his letters to Voltaire, writes that relating to the disaster of Lisbon, nature had not met in that place 20,000 houses of six or seven stories: "if the residents of this large city had been more evenly dispersed and less densely housed, the losses would have been fewer or perhaps none at all". Likewise, Kant (1756a: chapter 6) writes that we edify without giving too much weight to vaults, whose columns sometimes falter threatening to collapse.

From this particular point of view, the natural disaster is attributed to the non-observance of a rule. Mary Douglas (2003), in this respect, classifies institutional structures belonging to different cultures in accordance with three functional typologies: the first one sees natural disasters as result of violations of rules.

In this scenario, the "risk"[7] concept (Voss 2005; Japp 2000; Beck 1998; Luhmann 1991: 17) and, in general, prevention policies are fundamental conditions in the governance of disasters, in order to respond to catastrophes starting from their

[7]Generally, we could identify three typologies of risk management in the field of natural disasters: (a) Preventive risk management that seeks to prevent the enhancement of new risks; (b) Corrective risk management, on the basis of which existing risks are managed and consequently reduced; (c) Compensatory risk management, which tries to support the communities resilience, when residual risks cannot be reduced.

causes. This pillar is consequently associated to a more balanced concept of "catastrophism" than that emerging from the first pillar.

4.1 Political and Legal Sources About Prevention and Disaster Risk Governance

The process leading to the establishment of a new international framework for disaster risk reduction was started from the Hyogo Framework for Action (2005/2015). It informs and guides policymakers towards an effective disaster risk management, highlighting the prevention role in this field (UNISDR 2005).

The third world conference on disaster risk reduction (UNISDR 2015a), which replaced the Hyogo Framework, took place in Sendai from 14 to 18 march 2015. It is critical to the decisions about Sustainable Development Goals, because poorer countries are likely to be neglected. During this event, there was one disagreement about the responsibility topic: the focus was precisely on the "common but differentiated responsibilities" principle. Many developed countries thought this principle should not have been considered for disaster risk reduction. Developing countries, on the contrary, declared that this principle was necessary to sustainable development. The Sendai Conference delivered a document containing seven non-binding commitments. They shall aim to reduce impacts and increase resilience to risks by 2030, although the frequent use of the adverb "substantially" suggests a meaning of vagueness.

At this point, I recall legal[8] and meta-legal sources in the field of disasters governance: the Directive 2007/60/EC on floods, the Directive 2014/52/EU on the assessment of the effects of certain public and private environmental projects,

[8]Some cases decided by the European Court of Human Rights are important within this scenario. For instance, in the *Murillo Saldias and Others vs. Spain* case (28 November 2006) the applicants were survivors of the disaster which struck the Biescas campsite (Spanish Pyrenees) in 1996, where 87 people were killed in severe flooding following torrential rain. They complained in particular that Spain had not taken all the preventive measures that were necessary. However, the Court declared this application inadmissible. The *Budayeva and Others vs. Russia* case (20 March 2008) described indeed as the Republic of Kabardino-Balkariya (Russia) was devastated by a mudslide. The applicants alleged in particular that the Russian authorities had failed to mitigate the consequences of the mudslide. The Court, in fact, held that there had been a violation of Article 2 of the European Convention on Human Rights under its substantial and procedural limb. The *Özel and Others vs. Turkey* case (17 November 2015), instead, focused on the earthquake in Turkey (on 17 August 1999). The Court held that the Turkish authorities had not acted promptly in defining the responsibilities and causes of the collapse of the buildings which had determined the deaths. I also recall the *Viviani and Others vs. Italy* case (24 March 2015) concerning the risks attached to a potential eruption of Vesuvius and the measures taken by the authorities to combat those risks and, not at least, the *Kolyadenko and Others vs. Russia* case (28 February 2012). Furthermore, in this jurisprudential framework, un policy paper of the International Criminal Court (The Office of the Prosecutor, "Policy Paper on Case Selection and Prioritisation", 15 September 2016) seems rather significant. In effect, among the conducts which constitute a

the Dir. Seveso III 2012/18 EC on the control of major-accident hazards involving dangerous substances and the legislative proposals presented by the European Commission on integrating adaptation in the prevention and management of catastrophe risk [COM (2011) 934 final]. The legal sources, however, would seem rather limited and fragmented in respect to political sources.[9] Among these last sources, in addition to the Senday and Hyogo Conferences, I would mention first of all the European Parliament Resolution 2013/C56 E/04, which underlines the imperative need to integrate the policy of strengthening of response capacity into emergencies EU policies, the Proposal for the Resolution 2013/2683 about flood control and drainage infrastructure, the Resolution 2012/C 50 E/04 on the prevention of natural and man-made disasters, the Committee of the Regions Opinion 2013/C 218/07 on revision of the Directive concerning environmental impact assessment, the Communications of the European Commission aiming to strengthen the prevention and disaster risk management [COM (2008) 130 final,[10] COM (2009) 82 final,[11] COM (2009) 84,[12] COM (2010) 2020[13]] and the Decision 2007/779/EC, inherent in the European Civil Protection Mechanism. This latter was established at first by the European Union in 2001 about cases of disasters and it is activated through a structure for civil protection, characterized by a monitoring and information center within a common system of communication and emergency information. Civil protection assistance mainly consists of governmental aid delivered in the immediate aftermath of a disaster.[14]

The concluding document of the World Summit on Sustainable Development (United Nations 2002) provides the indication of a timely notification to other

---

serious crime under national law, the illegal exploitation of natural resources, the land grabbing and the environmental "destruction" will be widely considered.

[9]It is interesting to note that the European Directive on liability for environmental damage (Directive 2004/35/EC) does not cover the environmental damage caused by "a natural phenomenon of an exceptional, inevitable and uncontrollable character". Therefore, natural disasters do not fall within the implementation of this source.

[10]The Commission highlights the need for an integrated approach covering risk assessment, prediction, prevention, preparedness, rehabilitation and the opportunity to strengthen the relationship between civil protection and environmental policies.

[11]It suggests a European disasters prevention strategy with measures about the reduction of their frequency and limitation of the consequences.

[12]It concerns the countries that are beyond the EU's borders, in relation to which the Commission proposes a strategy for risk reduction in developing countries.

[13]In this Strategy for increasing smart, sustainable and inclusive Europe, (Europe 2020), the improvement of the strength of our economies to climate risks is required, as well as the ability to prevent and respond to disasters.

[14]The Mechanism currently includes all 28 EU Member States, in addition to Iceland, Montenegro, Norway, Serbia, the former Yugoslav Republic of Macedonia and Turkey. It intervened in some of the most devastating disasters the world has faced, like the earthquake in Haiti (2010), the triple-disaster in Japan (2011), typhoon Haiyan that hit the Philippines (2013), the floods in Serbia and Bosnia and Herzegovina (2014), the Ebola outbreak (2014), the conflict in Ukraine (2014), the earthquake in Nepal (2015) and the refugee crisis in Europe.

States of any natural disaster and other emergency situations too, likely to produce unanticipated adverse effects on the environment, also in order to facilitate international cooperation to support affected populations.

Even the World Health Assembly Resolution (number 61.19), adopted at the sixty-first World Health Assembly (2008) and entitled "Climate Change and Health", urges Member States to strengthen the monitoring of health systems, which can minimize the impact on public health, through adequate preventive measures (the preparation, a timely response and an effective management of the natural disasters).

A good governance of disasters also includes a sustainable adaptation policy (European Commission 2013a). The Action 8 of the European strategy on adaptation to climate change is an important source in this framework, since it promotes the insurance against disasters, as well as the United Nations Framework Convention on Climate Change, which clarifies the priorities for adaptation options, among which the management of extreme events and disasters emerges.[15]

The European Green Paper on Insurance for Natural Disasters (European Commission 2013b) reports that because of climate change, insurance will be solicited to cover more and more frequent and intense phenomena.[16] This source stresses the importance of 'insurance-linked securities' in relation to climate change. The Green Paper aims to improve the market for disasters insurance in the European Union. In this respect, the research undertaken by the EU Joint Research Centre (2011) underlines that flood, storm and earthquake risk is heterogeneous among EU Member States. However, we must recognise that there are cases where disaster insurance markets do not seem to cope fully with existing risks. Furthermore, the most vulnerable countries would face comparatively high premiums, which they presumably could not afford. In this scenario, we should also consider that 66% of the global population is expected to be living concentrated in cities by 2050, and that disasters prevention is often more effective in terms of cost, rather than fight the same disasters.[17]

---

[15]The mention of climate insurance is largely originating from The Alliance of Small Island States, which in 1991 proposed the creation of an international insurance device, in order to compensate damages from sea level rise. The section 4.8 of the UNFCCC, the article 3.14 of the Kyoto Protocol and the 2007 Bali Action Plan are further sources on this subject.

[16]Among international best practices in this area, I would like to mention the Mauritius Sugar Insurance Fund. It provides protection to the island's sugar farmers against losses from cyclones, fire, extreme rain and yellow spot disease (Linnerooth-Bayer and Mechler 2006).

[17]Just consider that between 1980 and 2011, the economic toll of natural disasters in the whole of Europe reached 445 billion euro (European Commission 2013b). Moreover, the European Environment Agency (2012) estimates that river flooding costs 6.4 billion euros annually. In accordance with the report by the United Nations Office for Disaster Risk Reduction (UNISDR 2015b), instead, economic losses from disasters are now averaging at US$250–300 billion each year. The annual investment in disaster risk reduction is estimated to account for only 0.1% of the US$6 trillion per year that will need to be invested in infrastructure alone over the upcoming 15 years (UNISDR 2015b).

5 Solidaristic-Transformative Pillar

This third and last pillar focuses on the “positive” aspects involving disasters. In keeping with this perspective, a catastrophe may represent an “opportunity for change”: the opportunity to begin again, to love, to discover the “advantages”, to cancel the differences, to show a “learning” outcome, to rebuild not only the urban and social fabrics, but also the spiritual and cognitive development.

Rousseau (1764), for example, states that the pessimism of Voltaire is cruel. The Alexander Pope’s Poem (1733) relieves his evils and invites him to the virtue of patience; on the contrary, Voltaire, with his negativity, leads him to despair. For Kant (1756a: chapter 6), instead, the damages that man may have suffered because of earthquakes are offset by more benefits. The view of the miseries inflicted by disasters, indeed, should result in “love for all men” (Kant 1756b: chapter 7).

Similarly, John Woodward interprets the biblical catastrophe as a universal dissolution, of educational, non-destructive significance[18] in his essay “Towards a Natural History of the Hearth” (Woodward 1695), while Günther Anders (1957: 11–12) believes that “the possibility of the Apocalypse is our work. But we know not what we are doing. We really don’t know, nor do they who control the Apocalypse”.

Along the same lines, Hans Jonas (1996) argues that the probability of ecological catastrophe gives “optimism” in relation to humanity’s capacity to produce a just “moral response”.

In Jean Pierre Dupuy’s opinion (2002), instead, human beings must take on a responsibility that is very similar to the concept of responsibility ideated by Hans Jonas (1979), but unlike this Author, Dupuy does not consider it necessary to review the relationship with nature, but rather he wants to deepen the social nature of technological design. The concept of “catastrophe” for Dupuy is wrongly linked to the idea of impotence in the face of nature. This concept, indeed, would be connected to the recognition of a power, first technique, that we would no longer be able to control. In the light of the thought of Günther Anders (1957) on nuclear and the Hans Jonas’ theory on the principle of responsibility (Jonas 1979), Dupuy suggests the idea of an imbalance between human technical power and man’s ability to foresee the consequences of his actions. He criticizes the precautionary principle, which would be steeped in a utilitarian logic, only capable of calculating the costs and benefits of actions. It is necessary that the disaster, which obviously we do not want, becomes a reality, so that we can plan a joint action. In his view, the idea of “time” should be revisited: it is appropriate, in fact, neglecting the awareness of a “historical time”, which from the past goes to the time in which many options are still open, that is the future time. On the contrary, it is opportune to accept the idea of a “circular time”, able to create a projectual circuit between past and future. If we put ourselves in this future reality of the occurred catastrophe, we

[18]The awareness of sin and the need for regeneration of the world make the idea of catastrophe, understood as an unquestioned and sensible traumatic change.

can live the present time in a different way, finding in it new possibilities that confirm that man is still in time to change his own future for the better and that of his descendants. In the words of this scholar, the human inability to believe in catastrophe[19] depends not only from a mistaken conception of time, but also from a process of desacralization of the world (Dupuy 2010).

The "enlightened catastrophism" suggested by Dupuy is not only a critique of the precautionary principle, but also a philosophical reversal of our ways of thinking about the world and time, based on the temporality of catastrophes. This catastrophism, in other words, is the theoretical attempt to recognize that the seemingly impossible—as a catastrophe—is possible and even certain. Nevertheless, we have the ability to operate responsibly to fight disasters. I fully agree with Dupuy when he says that we have acquired the means to destroy the planet and ourselves,[20] but we have not changed our way of thinking (Dupuy 2002).

The notion of "enlightened catastrophism" approaches to the "catastrophic change" ideated by Wilfred Ruprecht Bion (1965). He supports the concept of "catastrophe" to the concept of "change" in his monograph "Transformations" (Bion 1965). According to Bion, in fact, we find ourselves before a catastrophic change when a given event causes a subversion of the established order or of the system of things. "Catastrophic", because it is perceived emotionally as a disaster that impacts so suddenly and violently. Bion compares this change to an explosion that transforms a pre-catastrophic moment (characterized by the absence of emotions and an impoverishment of psychic life) in a post-catastrophic moment (marked by emotions, functional to psychic growth). Therefore, the "catastrophism" is for Bion an intense emotional situation that occurs in the process of psychic growth (Bion 1965).

Anselm Grün (2004), for his part, thinks that in the face of natural disasters we could assume three basic tasks: the first one would be to attempt to structure the world in a way, which is more respectful of nature. The second task is to review the way in which we see nature and creation, and to respect the power of the forces of nature (such as the power of lightning, thunder, earthquakes, volcanic eruptions and tsunamis). The third task, finally, refers to the whole image of God.

Even from the point of view of the Catholic faith, catastrophe can be seen as an "opportunity". God does not allow evil, but if He allows it, this is because from the same evil, He draws the good, in ways that we shall fully know only in eternal life.[21] A catastrophe can therefore be seen as an opportunity, provided to man in order to be "missionary", that is, to rediscover solidarity in bringing material or spiritual aid

---

[19]According to Dupuy, the economy and society suffer, as politicians, technicians and actors of world capitalism do not believe in the future, because they do not know what it means to create it. In this way, they are victims of the repetition of the present, and then victims of an "unconscious catastrophism".

[20]Directly, with weapons of mass destruction and, indirectly, through the change of the conditions necessary for human survival (Dupuy 2010).

[21]Catechism of the Catholic Church, n. 324.

to his neighbor.[22] Hence, the concept of "catastrophism" can therefore easily be approximated to the notions of "solidarity"[23] and "opportunity".

6 Towards the Conclusions: The "Responsible Catastrophism Model"

The "Responsible Catastrophism Model" is a new model that I want to suggest in the framework of disaster studies. It takes into account all three pillars described above.

With reference to the first pillar, the responsible catastrophism preserves the "awareness" of human powerlessness compared to the rebellion of nature. We often take for granted that human beings have a natural right to a long and safe life. In order to maximize this "awareness", people or society in general should think not only about the number of victims, but also about the moral possibility to actualize the concept of "time". This model also includes a broad concept of vulnerability affecting the spheres of fundamental rights (to life, survival, health, property, etc.). This vulnerability is functional to some factors, such as culture, systems of prevention, security and health, institutional stability, inequalities and life styles.

Relative to the second pillar, instead, the responsible catastrophism model accepts the idea of reducing risk and applying international, European and national public policies about prevention, but it also takes account of a new concept of risk, which considers a low chance of the disaster occurring. In the field of disaster studies, actually, the concept of "risk" is very important, but it is not enough, because catastrophic events are often characterized by weak probability, therefore high uncertainty, and hard impacts. In this case, I could also use the definition of "reticent risk" (or "silent risk"), understood as an existing risk, although weak, but which decides to manifest itself unexpectedly, causing severe impacts.

Finally, in connection with the third pillar, the responsible catastrophism model welcomes the optimism lying in the opportunity given by the disaster. However, this optimism should be anticipated at the current time, through a "positive correction" of our own behavior. Often we are accustomed to seeing only the negative side of things or situations of life. We should go back to look at life with the eyes of a child. This does not mean seeing in a childlike way, but with the inner eye of the new conscious opportunities.

[22]I recommend the book: "A Paradise Built in Hell: The Extraordinary Communities that Arise in Disaster" (Solnit 2009).

[23]John Paul II (2004) wrote in this regard that in our days human society seems shrouded in dark shadows, while it is shaken by tragic events and shattered by catastrophic natural disasters. However, "as on the night in which Jesus was betrayed (1 Cor. 11, 23), today He breaks the bread for us (Mt. 26, 26) in our Eucharistic celebrations and offers Himself under the sacramental sign of His Love for all. Therefore, the Eucharist is not only an expression of communion in the Church's life; it is also a project of 'solidarity' for all humanity" (Paul II 2004).

If we see "catastrophism" as a means of changing, as a tool to seize opportunities in order to become better people, then, the "integrity" perspective (Westra 2016) is consistent with this framework. By means of moral, ethical, spiritual, political, social and psychological tools, we have the opportunity to anticipate in our own lives the possibility of a catastrophic moment, so as to leave an open space to the change. Conversely, if "catastrophism" is interpreted as a negative scenario, where everything is bad and pessimism reigns and where a natural disaster is seen only as a means leading to deaths or other damages, and not also as an opportunity for a deeper reflection, the "disintegrity" dimension is dominant.

In conclusion, the "responsible catastrophism model" is able to identify a novel multilevel responsibility from ethical, social and moral matrix, where the concepts of "solidarity" and "survival" have a certain value.

References

Allen K (2003) Vulnerability reduction and the community-based approach. In Pelling M (ed) Natural disasters and development in a globalizing world, pp 170–184

Anders G (1957) Commandments in the atomic age. In: Burning conscience 1961. Monthly Review Press, New York, pp 11–12

Beck U (1998) World risk society. Polity Press, Cambridge

Bion WR (1965) Transformations, 1984. Karnac, London

Birkmann J, Cardona OD, Carreno ML, Barbat AH, Pelling M, Schneiderbauer S, Kienberger S, Keiler M, Alexander DE, Zeil P, Welle T (2014) Theoretical and conceptual framework for the assessment of vulnerability to natural hazards and climate change in Europe. In Birkmann J, Kienberger S, Alexander DE (eds) Assessment of vulnerability to natural hazards. A European perspective, chapter 1. Elsevier, Waltham, pp IX–XIII

Cutter SL (2003) The vulnerability of science and the science of vulnerability. Ann Assoc Am Geogr 93(1)

Cuvier BG (1827) Essay on the theory of the earth, 5th edn. T. Cadell, London

Development Assistance Research Associates and Climate Vulnerable Forum (2012) Climate vulnerability monitor: a guide to the cold calculus of a hot planet, 2nd edn. Madrid/Geneva

Douglas M (2003) Risk acceptability according to the social sciences. Routledge, London/New York

Dupuy JP (2002) Pour un catastrophisme éclairé – Quand l'impossible est certain. Seuil, Paris

Dupuy JP (2010) Le Souci de l'Avenir. La Revue Socialiste - Le dossier, n. 38

European Commission (2013a) Communication from the Commission to the European Parliament, the Council, the European Economic and Social Committee and the Committee of the Regions. An EU strategy on adaptation to climate change, COM (2013) 216 final, SWD(2013) 132 final

European Commission (2013b) Green paper on the insurance of natural and man-made disasters: COM/2013/0213 final

European Environment Agency (2012) Climate change, impacts and vulnerability in Europe 2012. An indicator-based report, EEA Report No 12/2012

Ford JD, Berrang-Ford L, King M, Furgal C (2010) Vulnerability of aboriginal health systems in canada to climate change. Glob Environ Chang 20

Füssel HM (2007) Vulnerability: a generally applicable conceptual framework for climate change research. Glob Environ Chang 17:155–167

Global Humanitarian Forum (2009) Human impact report. Climate change: the anatomy of a silent crisis. Global Humanitarian Forum, Geneva
Grün A (2004) Womit Habe Ich Das Verdient? - Die Unverständliche Gerechtigkeit Gottes. Vier Türme Verlag
Hutton J (1785) Theory of the earth. An investigation of the laws observable in the compositions. Dissolution and restoration of land upon the globe. Transactions of the Royal Society of Edinburgh
IPCC (2001) Climate change 2001: impacts, adaptation and vulnerability – summary for policymakers. Cambridge Universiy Press, Cambridge
IPCC (2007) In: Core Writing Team, Pachauri RK, Reisinger A (eds) Climate change 2007: synthesis report. Contribution of Working Groups I, II and III to the fourth assessment report of the intergovernmental panel on climate change. IPCC, Geneva
IPCC (2014) Fifth assessment report: climate change 2014: impacts, adaptation, and vulnerability. Cambridge University Press, Cambridge/New York
Japp KP (2000) Risiko. Transcript Verlag, Bielefeld
Joint Research Centre, European Commission (2011) Natural catastrophes: risk relevance and insurance coverage in the EU. Publications Office of the European Union, Luxembourg
Jonas H (1979) Das Prinzip Verantwortung: Versuch einer Ethik für die technologische Zivilisation, 1984. Neuauflage als Suhrkamp Taschenbuch, Frankfurt/M
Jonas H (1996) Toward an ontological grounding of an ethics of the future. In: Vogel L (ed) Mortality and morality: a search for the good after auschwitz. Northwestern University Press, Evanston, pp 99–112
Kant I (1756a) On the causes of earthquakes on the occasion of the calamity that befell the western countries of Europe towards the end of last year, chapter 6. In: Natural science, pp 327–336
Kant I (1756b) History and natural description of the most noteworthy occurrences of the earthquake that struck a large part of the earth at the end of the year 1755, chapter 7. In: Natural science, pp 337–364
Kelly PM, Adger WN (2000) Theory and practice in assessing vulnerability to climate change and facilitating adaptation. Clim Chang 47:325–352
Linnerooth-Bayer J, Mechler R (2006) Insurance for Assisting adaptation to climate change in developing countries: a proposed strategy, climate change and insurance. Clim Policy 6(6)
Luhmann N (1991) Soziologie des Risikos. Walter de Gruyter, Berlin/New York
Lyell C (1830) Principles of geology, being an attempt to explain the former changes of the earth's surface, by reference to causes now in operation. John Murray/Albemarle Street, London
Morzaria-Luna HN, Turk-Boyer P, Moreno-Baez M (2014) Social indicators of vulnerability for fishing communities in the Northern Gulf of California, Mexico: implications for climate change. Mar Policy 45
Myers CA, Slack T, Singelmann J (2008) Social vulnerability and migration in the wake of disaster: the case of Hurricanes Katrina and Rita. Popul Environ 29(6)
Naumann G, Barbosa P, Garrote L, Iglesias A, Vogt J (2014) Exploring drought vulnerability in Africa: an indicator based analysis to be used in early warning systems. Hydrolo Earth Syst Sci 18:1591–1604
Nel P, Righarts M (2008) Natural disasters and the risk of violent civil conflict. Int Stud Q 52:159–185
Paul II J (2004) Mane Nobiscum, Domine, 27. Vatican City
Pope A (1733) An essay on man. In: Epistles to a friend. J. Wilford, London
Rousseau JJ (1764) Lettre de J.J. Rousseau citoyen de Geneve, à Monsieur de Voltaire [18 août 1756] concernant le Poème sur le désastre de Lisbonne, 2014. Nabu Press, Paris
Schilling J, Freier KP, Hertige E, Scheffrana J (2012) Climate change, vulnerability and adaptation in North Africa with focus on Morocco. Agric Ecosyst Environ 156:12–26
Slettebak R & De Soysa I. (2010) High temps, high tempers? Weather-related natural disasters & civil conflict. Draft paper for the conference on climate change and security, Trondheim, Norway

Solnit R (2009) A paradise built in hell: the extraordinary communities that arise in disaster. The Viking Press, New York
Tierney K (2006) Social inequality, hazards and disasters. In: Daniels RJ, Kettl DF, Kunreuther H (eds) On risk and disaster: lessons from Hurricane Katrina. University of Pennsylvania Press, Philadelphia
Turner BLI, Kasperson R, Matson P, McCarthy JJ, Corell R, Christensen L, Eckley N, Kasperson JX, Luers AL, Martello ML et al (2003) A framework for vulnerability analysis in sustainability science. Proc Nat Acad Sci U S A 100
UNISDR - United Nations Office for Disaster Risk Reduction (2005) Hyogo framework for action 2005–2015: building the resilience of nations and communities to disasters. UN/ISDR, Geneva
UNISDR - United Nations Office for Disaster Risk Reduction (2015a) Sendai framework for disaster risk reduction 2015–2030. UN/ISDR, Geneva
UNISDR - United Nations Office for Disaster Risk Reduction (2015b) Making development sustainable: the future of disaster risk management. Global assessment report on disaster risk reduction. UN/ISDR, Geneva
United Nations (2002) Report of the world summit on sustainable development, Johannesburg/New York, 26 August–4 September
Vincent K (2004) Creating an index of social vulnerability to climate change for Africa. Centre for Climate Change Research. Working Paper, 56
Voltaire (Arouet F.M.) (1756) Poème sur le désastre de Lisbonne, 2011. Arvensa Edition, Paris
Voss M (2005) Towards a reopening of risk societies. A contribution to the debate on risk and danger. In: Learning about risk. Launch conference ESRC, social contexts and responses to risk, Canterbury, 28–29 January 2005
Weichselgartner J (2001) Disaster mitigation: the concept of vulnerability revisited. Disaster Prev Manag 10
Westra L (2016) Ecological integrity and global governance. Science, ethics and the law. Earthscan, London/Routledge, New York
WHO (2009) Global health risks: mortality and burden of disease attributable to selected major risks. World Health Organization, Geneva
Wisner B, Blaikie P, Cannon T, Davis I (2003) At risk: natural hazards, people's vulnelity and disasters. Routledge, London/York
Woodward J (1695) Towards a natural history of the hearth, 2014. Cambridge University Press, Cambridge
World Health Assembly (2008) Resolution WHA.61.19. Climate change and health, 24 May 2008

MIX
Papier aus verantwortungsvollen Quellen
Paper from responsible sources
FSC® C105338

If you have any concerns about our products,
you can contact us on
ProductSafety@springernature.com

In case Publisher is established outside the EU,
the EU authorized representative is:
Springer Nature Customer Service Center GmbH
Europaplatz 3, 69115 Heidelberg, Germany

Printed by Libri Plureos GmbH
in Hamburg, Germany